# WORKSHEETS
## FOR CLASSROOM OR LAB PRACTICE

CARRIE GREEN

# ELEMENTARY ALGEBRA:
# CONCEPTS AND APPLICATIONS
## EIGHTH EDITION

## Marvin L. Bittinger

*Indiana University Purdue University Indianapolis*

## David J. Ellenbogen

*Community College of Vermont*

**Addison-Wesley**
is an imprint of

PEARSON

Reproduced by Pearson Addison-Wesley from electronic files supplied by the author.

Copyright © 2010 Pearson Education, Inc.
Publishing as Pearson Addison-Wesley, 75 Arlington Street, Boston, MA 02116.

ISBN-13: 978-0-321-59931-5
ISBN-10: 0-321-59931-4

1 2 3 4 5 6 BB 11 10 09 08

**Addison-Wesley**
is an imprint of

www.pearsonhighered.com

# Table of Contents

# Chapter 1 INTRODUCTION TO ALGEBRAIC EXPRESSIONS

## 1.1    Introduction to Algebra

**Topics**
> Algebraic Expressions
> Translating to Algebraic Expressions
> Translating to Equations

**Key Terms**

Use the vocabulary terms listed below to complete each statement in Exercises 1–3.

           **constant**               **equation**            **variable**

1. In the expression $4x^3$, 4 is a(n) _____.

2. In the expression $4x^3$, $x$ is a(n) _____.

3. A(n) _____ is a mathematical sentence with expressions on each side of an equals sign.

*Evaluate.*

4. $5c$, for $c = 8$                             **4.**_____

5. $m - 7$, for $m = 16$                     **5.**_____

6. $\dfrac{a+b}{8}$, for $a = 14$ and $b = 18$      **6.**_____

7. $\dfrac{7c}{d}$, for $c = 9$ and $d = 3$           **7.**_____

**8.**  $\dfrac{x - y}{5}$, for $x = 19$ and $y = 14$

8._____

*Substitute to find the value of each expression.*

**9.** A driver who drives at a speed of $r$ mph for $t$ hr will travel a distance $d$ mi given by $d = rt$ mi. How far will a driver travel at a speed of 62 mph for 6 hr?

9._____

**10.** The area $A$ of a parallelogram with base $b$ and height $h$ is given by $A = bh$. Find the area of a parallelogram with a height of 11 cm and a base of 8 cm.

10._____

*Translate to an algebraic expression.*

**11.** 46 more than $t$

11._____

**12.** 19 less than $d$

12._____

**13.** 12 increased by $y$

13._____

**14.** The product of 6 and $c$

14._____

**15.** 4 divided by $z$

15._____

**16.**  $x$ subtracted from $q$                    **16.**_____

**17.**  The sum of $s$ and $d$                    **17.**_____

**18.**  A number $x$ plus 8 times $n$            **18.**_____

**19.**  One half of some number                   **19.**_____

**20.**  12% of $n$                                **20.**_____

*Determine whether the given number is a solution of the given equation.*

**21.**  22; $x + 12 = 37$                         **21.**_____

**22.**  10; $\frac{x}{7} = 5$                      **22.**_____

**23.**  2; $\frac{32}{x} = 16$                     **23.**_____

*Translate each problem to an equation. Do not solve.*

**24.** 93 minus what number is 48?                          **24.**_____

**25.** When 16 is multiplied by a number, the result       **25.**_____
is 128. Find the number.

**26.** A gameboard has 60 squares. If you win 25           **26.**_____
squares and your opponent wins the rest, how
many does your opponent win?

*In Exercises 27–30, match the phrase or sentence with the appropriate expression or equation.*

**27.** Twice the sum            **A.** $\frac{1}{2} \cdot u \cdot v$        **27.**_____

of two numbers

**28.** One less than twice      **B.** $2x - 1 = 5$        **28.**_____
a number is five.

**29.** Half the product         **C.** $m + 2 = 5$         **29.**_____
of two numbers

**30.** Two more than a          **D.** $2(c + d)$          **30.**_____
number is five.

# Chapter 1 INTRODUCTION TO ALGEBRAIC EXPRESSIONS

## 1.2     The Commutative, Associative, and Distributive Laws

**Topics**
   Equivalent Expressions
   The Commutative Laws
   The Associative Laws
   The Distributive Law
   The Distributive Law and Factoring

**Key Terms**
Use the vocabulary terms listed below to complete each statement in Exercises 1–3.

   **commutative**              **associative**              **distributive**

1.  The _____ law for multiplication states that changing the order
    of multiplication does not affect the answer.

2.  The law that enables you to write the product of a number and a sum as the sum of two
    products is the _____ law.

3.  The _____ law of addition applies when regrouping numbers to be
    added.

*Use the commutative law of addition to write an equivalent expression.*

4.  $18 + x$                                      4._____

5.  $9y + 4x$                                     5._____

*Use the commutative law of multiplication to write an equivalent expression.*

6.  $5z$                                          6._____

7.  $g + vu$                                      7._____

*Use the associative law of addition to write an equivalent expression.*

**8.** $z + (y + 93)$                     **8.** _____

**9.** $(b + cd) + 60$                    **9.** _____

*Use the associative law of multiplication to write an equivalent expression.*

**10.** $28 \cdot (v \cdot u)$                   **10.** _____

**11.** $2[3(a + b)]$                      **11.** _____

*Use the commutative and/or the associative laws to write two equivalent expressions. Answers may vary.*

**12.** $(x + y) + 15$                     **12.** _____

                                          _____

**13.** $65 \cdot (s \cdot r)$                     **13.** _____

                                          _____

*Use the commutative and/or associative laws to show why the expression on the left is equivalent to the expression on the right. Write a series of steps with labels. See Example 4 in the text.*

**14.** $5 + (9 + a)$ is equivalent to $a + 14$     **14.** _____

                                          _____

                                          _____

*Multiply.*

**15.**   $17(y + 6)$

**15.** _____

**16.**   $18(p + 1)$

**16.** _____

**17.**   $7(4q + 10)$

**17.** _____

**18.**   $4(m + 2 + 8n)$

**18.** _____

**19.**   $(r + n)6$

**19.** _____

**20.**   $(a + y + 6)4$

**20.** _____

*List the terms in each expression.*

**21.**   $g + ghj + 22$

**21.** _____

**22.**   $3m + \dfrac{m}{n} + 4n$

**22.** _____

*Use the distributive law to factor each of the following. Check by multiplying.*

**23.** $9x + 9y$

**23.**_____

**24.** $57x + 57$

**24.**_____

**25.** $30r + 42s$

**25.**_____

**26.** $2a + 2b + 20$

**26.**_____

*List the factors in each expression.*

**27.** $4s$

**27.**_____

**28.** $2(a + d)$

**28.**_____

**29.** $z \cdot 8$

**29.**_____

**30.** $(x + y)(x - y)$

**30.**_____

# Chapter 1 INTRODUCTION TO ALGEBRAIC EXPRESSIONS

## 1.3    Fraction Notation

**Topics**

Factors and Prime Factorizations
Fraction Notation
Multiplication, Division, and Simplification
More Simplifying
Addition and Subtraction

**Key Terms**
Use the vocabulary terms listed below to complete each statement in Exercises 1–4.

**prime          natural          numerator    denominator**

1.  In fraction notation, the number on top is called the_____.

2.  A _____ number has exactly two factors, 1 and itself.

3.  Another name for the counting numbers is the _____ numbers.

4.  To add fractions, they must have the same _____.

*Write all two-factor factorizations of each number. Then list all the factors of the number.*

5.  175                                        **5.**_____

_____

_____

6.  18                                         **6.**_____

_____

_____

*Find the prime factorization of each number. If the number is prime, state this.*

7.  187                                        **7.**_____

8.  225                                        **8.**_____

*Simplify.*

**9.** $\dfrac{40}{48}$

9._____

**10.** $\dfrac{450}{50}$

10._____

**11.** $\dfrac{3}{12}$

11._____

**12.** $\dfrac{15}{21}$

12._____

*Perform the indicated operation and, if possible, simplify. Check using a calculator.*

**13.** $\dfrac{2}{11} \cdot \dfrac{2}{13}$

13._____

**14.** $\dfrac{5}{10} + \dfrac{3}{10}$

14._____

**15.** $\dfrac{4}{9} - \dfrac{5}{27}$

15._____

**16.** $\dfrac{7}{5} \div \dfrac{13}{2}$

16._____

**17.** $\dfrac{3}{14} + \dfrac{4}{35}$

17._____

# Chapter 1 INTRODUCTION TO ALGEBRAIC EXPRESSIONS

## 1.4     Positive and Negative Real Numbers

| Topics |
| --- |
| The Integers<br>The Rational Numbers<br>Real Numbers and Order<br>Absolute Value |

**Key Terms**

Use the vocabulary terms listed below to complete each statement in Exercises 1–4.

       **rational**             **irrational**           **integer**           **inequality**

1. When written in decimal form, _____ numbers neither terminate nor repeat.

2. A(n) _____ is a whole number or its opposite.

3. The statement $4 < 9$ is an example of an _____.

4. A fraction of integers is a(n) _____ number.

*State which real numbers correspond to each situation.*

5. During a quiz show, a person loses 2100 points and then wins 4100 points.       5._____

                                                         _____

6. During a year, a person saves 1350 dollars and spends 625 dollars.       6._____

                                                         _____

7. While playing a video game, Cindy intercepted a missile worth 70 points, lost a starship worth 175 points, and captured a base worth 900 points.       7._____

                                                         _____

8. During one month, a person withdraws 235 dollars and deposits 175 dollars.       8._____

                                                         _____

*Graph each rational number on a number line.*

9. $-1.4$                                                9.

                                                      $\longleftarrow\!\!\!\!\!\!\!\!\!\!\!\!\!\!\!\!\!\!\!\!\!\!\!\!\!\!\!\!\!\!\!\!\!\!\!\!\!\!\!\!\!\!\!\!\!\!\!\!\!\!\!\!\!\!\!\!\!\!\!\!\!\!\!\!\!\!\!\!\!\!\!\!\!\!\!\!\!\!\!\!\rightarrow$
                                                               0

**10.** $\dfrac{13}{4}$

**10.**

<--------------------|--------------------->
                     0

*Write decimal notation for each number.*

**11.** $\dfrac{13}{11}$

**11.**_____

**12.** $\dfrac{4}{3}$

**12.**_____

**13.** $-\dfrac{5}{8}$

**13.**_____

**14.** $\dfrac{12}{11}$

**14.**_____

*Write a true sentence using either < or >.*

**15.** $-5 \;\square\; 6$

**15.**_____

**16.** $2 \;\square\; -15$

**16.**_____

**17.** $-9 \;\square\; 8$

**17.**_____

**18.** $-1 \;\square\; -7$

**18.**_____

**19.** $\dfrac{10}{17} \;\square\; \dfrac{3}{13}$

**19.**_____

*Write a second inequality with the same meaning.*

**20.** $-9 > t$

20._____

*Classify each inequality as either true or false.*

**21.** $5 \geq -8$

21._____

**22.** $-19 \leq 10$

22._____

*Find each absolute value.*

**23.** $|-26|$

23._____

**24.** $|18|$

24._____

**25.** $|-5.8|$

25._____

**26.** $\left|-\dfrac{6}{13}\right|$

26._____

**27.** $\left|\dfrac{0}{14}\right|$

27._____

*For Exercises 28–30, consider the following list:*

$$6.12, \ -4.7, \ 12, \ \frac{1}{2}, \ \frac{4}{9}, \ -\frac{1}{6}, \ -3, \ 1\frac{2}{3}, \ \frac{7}{5}, \ \sqrt{7}$$

**28.** List all rational numbers.

28._____

**29.** List all integers.

29._____

**30.** List all real numbers.

30._____

# Chapter 1 INTRODUCTION TO ALGEBRAIC EXPRESSIONS

## 1.5     Addition of Real Numbers

| Topics |
|---|
| Adding with the Number Line<br>Adding without the Number Line<br>Problem Solving<br>Combining Like Terms |

*In each of Exercises 1–4, match the term with a like term from the column on the right.*

**1.** 49          **A.** $-3z$                    1._____

**2.** 5z          **B.** $17t$                    2._____

**3.** 4m          **C.** $m$                      3._____

**4.** $-2t$       **D.** $-22$                    4._____

*Add using the number line.*
**5.** $1+(-3)$                                     5._____

**6.** $-10+9$
                                                    6._____

*Add. Do not use the number line except as a check.*
**7.** $-3+3$                                       7._____

**8.** $10+(-4)$                                    8._____

**9.** $0 + (-99)$

_____

**10.** $-84 + (-4)$

**10.**_____

**11.** $-43 + 91$

**11.**_____

**12.** $96 + (-52)$

**12.**_____

**13.** $-69 + (-41)$

**13.**_____

**14.** $-11.6 + 31.8$

**14.**_____

**15.** $-77.2 + (-10.3)$

**15.**_____

**16.** $-\dfrac{5}{6} + \dfrac{10}{3}$

**16.**_____

**17.** $-\dfrac{7}{10} + \left(-\dfrac{9}{30}\right)$

**17.**_____

**18.** $-\dfrac{9}{14} + \dfrac{3}{7}$

**18.**_____

**19.** $4+(-12)+(-77)+(-63)$

**19.**_____

**20.** $-97.3+(-49.6)+24.8$

**20.**_____

**21.** $-25.3+(-31.5)+30.4$

**21.**_____

*Solve. Write your answer as a complete sentence.*

**22.** The barometric pressure at a certain city dropped 4 millibars (mb); then it rose 2 mb. After that, it dropped 11 mb and then it rose 9 mb. What was the total change in pressure?

**22.**_____

**23.** Kyle's credit card bill is $514. Kyle sends a check to the credit card company for $81, charges another $174 in merchandise, and then pays off another $200 of the bill. How much does Kyle owe the company?

**23.**_____

**24.** A mountain with a base 5,779 feet below sea level rises 17,780 feet. What is the elevation above sea level of its peak?

**24.**_____

*Combine like terms.*

**25.** $6z + 3z$

**25.**_____

**26.** $-1z + 3z$

**26.**_____

**27.** $-4b + (-2b)$

**27.**_____

**28.** $-7 + 7x + 9 + (-11x)$

**28.**_____

*Find the perimeter of each figure.*

**29.**

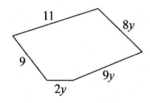

**29.** _____

**30.**

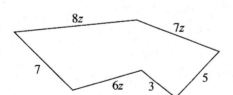

**30.** _____

# Chapter 1 INTRODUCTION TO ALGEBRAIC EXPRESSIONS

## 1.6    Subtraction of Real Numbers

| Topics |
| --- |
| Opposites and Additive Inverses |
| Subtraction |
| Problem Solving |

In each of Exercises 1–2, choose the appropriate wording for each expression.

**1.** $-b-8$                                                                    1._____

**A)** $b$ minus eight                  **B)** negative $b$ minus eight

**C)** negative $b$ minus negative eight   **D)** eight minus $b$

**2.** $14-(-n)$                                                                  2._____

**A)** $n$ minus 14                     **B)** fourteen minus $n$

**C)** fourteen minus negative $n$      **D)** negative $n$ minus fourteen

*Write each of the following in words*

**3.** $c-(-7)$                                                                   3._____

**4.** $-9-d$                                                                     4._____

**5.** Find the opposite, or additive inverse, of 66.          5._____

**6.** Find $-x$ when $x = -10.2$.                             6._____

**7.** Find $-(-x)$ when $x = -12$.                            7._____

**8.** Find the opposite of $-35$.                             8._____

*Subtract.*

**9.** $55 - 65$

**9.** _____

**10.** $80 - 2$

**10.** _____

**11.** $-86 - 12$

**11.** _____

**12.** $-11 - (-11)$

**12.** _____

**13.** $98 - 88$

**13.** _____

**14.** $-13 - 41$

**14.** _____

**15.** $8 - 42$

**15.** _____

**16.** $-52 - 98$

**16.** _____

**17.** $-4 - 8$

**17.** _____

**18.** $40.3 - 77$

**18.** _____

**19.** $-\dfrac{4}{5} - \dfrac{2}{3}$

**19.** _____

**20.** $\dfrac{14}{13} - \dfrac{11}{13}$

**20.** _____

**21.** $-\dfrac{5}{6} - \dfrac{4}{3}$

**21.** _____

*In each of Exercises 22–23, translate the phrase to mathematical language and simplify.*

**22.** The difference between 2.5 and −2.3

**22.** _____

**23.** The difference between −68 and 118

**23.** _____

*Simplify.*

**24.** $60 - (-87) - 6 - (-19) + 77$

**24.** _____

**25.** $45 - (-47) - 85 - (-93) + 55$

**25.** _____

**26.** Identify the terms in the expression.
$2y - 14z - 14$

**26.** _____

*Combine like terms.*
**27.** $14a + 13 - 18a$                    **27.**_____

**28.** $2 - 9y - 4 - 3y$                    **28.**_____

**29.** $-10 + 2a + 5b - 82 + b - 7a$         **29.**_____

*Solve.*
**30.** A submarine at a depth of 1521 ft ascends to a         **30.** _____
depth of 596 ft. How far did the submarine ascend?

# Chapter 1 INTRODUCTION TO ALGEBRAIC EXPRESSIONS

## 1.7    Multiplication and Division of Real Numbers

| Topics |
|---|
| Multiplication |
| Division |

*Multiply.*

**1.**  $(63)(-47)$                          **1.**_____

**2.**  $(-13)(-4)$                          **2.**_____

**3.**  $(90)(60)$                           **3.**_____

**4.**  $(-81)(77)$                          **4.**_____

**5.**  $8.2(29)$                            **5.**_____

**6.**  $17 \cdot 98$                        **6.**_____

**7.**  $-8 \cdot (-6.1)$                     **7.**_____

**8.**  $-\dfrac{8}{7} \cdot \left(-\dfrac{3}{1}\right)$          **8.**_____

**9.**  $17 \cdot (39) \cdot (-10) \cdot (13)$           **9.**_____

*Divide, if possible, and check. If a quotient is undefined, state this.*

**10.** $(-12) \div (6)$

**10.** _____

**11.** $\dfrac{20}{10}$

**11.** _____

**12.** $\dfrac{-132}{11}$

**12.** _____

**13.** $5 \div 0$

**13.** _____

**14.** $-14.4 \div 1.8$

**14.** _____

*Write each number in two equivalent forms.*

**15.** $\dfrac{-10}{7}$

**15.** _____

**16.** $-\dfrac{2}{9}$

**16.** _____

*Find the reciprocal of each number, if it exists.*

**17.** $\dfrac{-7}{5}$

**17.** _____

**18.** $8$

**18.** _____

**19.** $1.2$

**19.** _____

*Perform the indicated operation and, if possible, simplify. If a quotient is undefined, state this.*

**20.** $-\dfrac{6}{7} \cdot \left( \dfrac{3}{5} \right)$

                                                   **20.**_____

**21.** $-\dfrac{4}{3} \cdot \left( -\dfrac{5}{6} \right)$

                                                   **21.**_____

**22.** $-\dfrac{9}{5} + \left( -\dfrac{4}{5} \right)$

                                                   **22.**_____

**23.** $-\dfrac{3}{5} \div \left( -\dfrac{8}{7} \right)$

                                                   **23.**_____

**24.** $-\dfrac{2}{5} - \dfrac{3}{2}$

                                                   **24.**_____

**25.** $-\dfrac{1}{4} + \dfrac{5}{2}$

                                                   **25.**_____

**26.** $-\dfrac{4}{7} \div \dfrac{8}{5}$

                                                   **26.**_____

**27.** $-\dfrac{2}{5} \div \dfrac{9}{7}$

                                                   **27.**_____

**28.** $-\dfrac{5}{6}-\dfrac{2}{5}$                    **28.** _____

**29.** $-42.6 \div 21.3$                    **29.** _____

**30.** $-\dfrac{9}{14}+\dfrac{5}{7}$                    **30.** _____

# Chapter 1 INTRODUCTION TO ALGEBRAIC EXPRESSIONS

## 1.8    Exponential Notation and Order of Operations

> **Topics**
> Exponential Notation
> Order of Operations
> Simplifying and the Distributive Law
> The Opposite of a Sum

*Write exponential notation.*

**1.** $t \cdot t \cdot t$

1._____

**2.** $4y \cdot 4y \cdot 4y \cdot 4y \cdot 4y$

2._____

*Simplify.*

**3.** $(-5)^3$

3._____

**4.** $7^4$

4._____

**5.** $12^1$

5._____

**6.** $(6y)^2$

6._____

**7.** $99 - 14 \cdot 7$

7._____

**8.** $5 \cdot 3 + 7 \cdot 6$

8._____

**9.** $6^3 \div 36 - 6$

**9.** _____

**10.** $33 + 9^3 \div (-72) \cdot 8$

**10.** _____

**11.** $(-60) \div (-5) \cdot \left(-\dfrac{1}{3}\right)$

**11.** _____

**12.** $8(-3) + |2(-6)|$

**12.** _____

**13.** $\dfrac{3 \cdot 3 - 2^2}{256 - 4^3}$

**13.** _____

**14.** $(-3)^3$

**14.** _____

**15.** $14 - 7 \cdot 3 + 8$

**15.** _____

**16.** $128 \div (-4)^3 + 2\left[3 - 6(9-5)^2\right]$

**16.** _____

**17.** $\dfrac{(-5)^3 + 3^2}{4 \cdot 5 - 6^2 + 2 \cdot 7}$

**17.** _____

*Evaluate.*

**18.** $3x \div 6x^2$, for $x = 2$

**18.** _____

**19.** $-m^2 - 5m$, for $m = -7$

**19.** _____

**20.** $(75 \div x^2) - 7(x - 7)$, for $x = -5$

**20.** _____

**21.** $\dfrac{2a - a^2}{a^2 - 8}$, for $a = -3$

**21.** _____

*Write an equivalent expression without using grouping symbols.*

**22.** $-(s - 68)$

**22.** _____

**23.** $-(29c - 86d + 37)$

**23.** _____

**24.** $-\left(-15x + 6y - 8\right)$

**24.** _____

*Simplify.*

**25.** $15z - (9z - 29)$

**25.**_____

**26.** $46z - 53 - (57 + 37z)$

**26.**_____

**27.** $9c + 2d - 5(6c - 7d)$

**27.**_____

**28.** $6h + 6j - 5(6h - 7j + 5k)$

**28.**_____

**29.** $7n^4 + 3 - (6n^4 + 1)$

**29.**_____

**30.** $6(-2x - 5) - [3(x - 7) + 4]$

**30.**_____

# Chapter 2  EQUATIONS, INEQUALITIES, AND PROBLEM SOLVING

## 2.1    Solving Equations

**Topics**
    Equations and Solutions
    The Addition Principle
    The Multiplication Principle
    Selecting the Correct Approach

*Solve using the addition principle.*

**1.**  $x + 8 = 50$                                    **1.**_____

**2.**  $-34 = x - 16$                                 **2.**_____

**3.**  $-25 = x - 18$                                 **3.**_____

**4.**  $3 + x = 12$                                    **4.**_____

**5.**  $-16 + p = 31$                                 **5.**_____

**6.**  $x + \dfrac{2}{11} = \dfrac{7}{11}$            **6.**_____

**7.** $x - \dfrac{5}{6} = \dfrac{7}{8}$

7._____

**8.** $6.4 = x + 3.7$

8._____

**9.** $-5.4 = -2.5 + x$

9._____

**10.** $x + \dfrac{1}{10} = -\dfrac{3}{5}$

10._____

*Solve using the multiplication principle.*
**11.** $7x = 42$

11._____

**12.** $36 = 6x$

12._____

**13.** $-x = 31$

13._____

**14.** $-6.4x = 64$

14._____

**15.** $\dfrac{x}{4} = -10$

15._____

**16.** $\dfrac{2}{13}x = 8$

16._____

**17.** $-\dfrac{x}{5} = \dfrac{1}{8}$

17._____

**18.** $-\dfrac{2}{7}x = \dfrac{8}{35}$

18._____

**19.** $-\dfrac{1}{2}x = -\dfrac{9}{10}$

19._____

**20.** $-\dfrac{x}{7} = \dfrac{1}{16}$

20._____

*Solve.*

**21.** $105 = s - 7.3$

21._____

**22.** $39.7r = 238.2$

22. _____

**23.** $-49 + m = 13$

23. _____

**24.** $-\dfrac{5}{13} + z = \dfrac{-10}{9}$

24._____

**25.** $\dfrac{-5}{9}p = -15$

25._____

**26.** $\dfrac{3}{10}n = 9$

26._____

**27.** $8.4 = x - 8.5$

27._____

**28.** $-46 + t = 35$

28._____

# Chapter 2  EQUATIONS, INEQUALITIES, AND PROBLEM SOLVING

## 2.2    Using the Principles Together

| Topics |
| --- |
| Applying Both Principles |
| Combining Like Terms |
| Clearing Fractions and Decimals |

*Solve and check. Label any contradictions or identities.*

**1.**  $7x + 8 = 15$                          1._____

**2.**  $9x - 4 = 77$                          2._____

**3.**  $-103 = 5 + 9x$                        3._____

**4.**  $5x + 7x = 72$                         4._____

**5.**  $-4x - 13x = 119$                      5._____

**6.**  $10.8x - 4.9x = -177$                  6._____

**7.**  $x + \dfrac{1}{2}x = 12$               7._____

**8.** $9x - 20 = 5x$                    **8.**_____

**9.** $4y - 2 = 33 - 3y$               **9.**_____

**10.** $7 + 2s - 6 = 6s + 3 - 3s$      **10.**_____

**11.** $6 - 2x = 6x - 10x + 16$        **11.**_____

**12.** $3(2t - 1) = 9$                 **12.**_____

**13.** $24 = 6(x - 5)$                 **13.**_____

**14.** $2(z - 4) = 4(z + 1)$           **14.**_____

**15.** $6x - x = 3x + x$               **15.**_____

**16.** $7 + 3s - 5 = 8s + 6 - 4s$

16._____

**17.** $6(x - 2) = 5(x + 2) + 2x$

17._____

**18.** $x + \dfrac{1}{6}x = 14$

18._____

*Clear fractions or decimals, solve, and check.*

**19.** $\dfrac{5}{8}x + \dfrac{1}{16}x = \dfrac{9}{16} + x$

19._____

**20.** $-\dfrac{5}{3}x + \dfrac{2}{3} = -16$

20._____

**21.** $\dfrac{8}{3} + \dfrac{1}{6}y = \dfrac{4}{18} - \dfrac{4}{6}$

21._____

**22.** $2.1x - 1.72 = 0.84 - 4.3x$

22._____

**23.** $\dfrac{5}{2}x + \dfrac{1}{4}x = \dfrac{9}{4} + x$

**23.** _____

**24.** $\dfrac{1}{3}(3y + 12) - 6 = -\dfrac{1}{3}(6y - 21)$

**24.** _____

**25.** $\dfrac{2}{3}\left(\dfrac{5}{8} + 8x\right) - \dfrac{3}{8} = \dfrac{5}{8}$

**25.** _____

**26.** $0.7(5x + 3) = 1.1 - (x + 5)$

**26.** _____

**27.** $1 - \dfrac{2}{7}t = \dfrac{5}{7} - \dfrac{3}{7}t + \dfrac{2}{3}$

**27.** _____

**28.** $0.6 - 3(c - 2) = 0.1 + 4(5 - c)$

**28.** _____

# Chapter 2  EQUATIONS, INEQUALITIES, AND PROBLEM SOLVING

## 2.3    Formulas

**Topics**
  Evaluating Formulas
  Solving for a Variable

**Key Terms**
Use the vocabulary terms listed below to complete each statement in Exercises 1–2.

**circumference**                    **formula**

1.  A _____ is an equation relating several quantities.

2.  The _____ is the distance around a circle.

*Solve.*

3.  The formula $d = 65t$ gives the distance                3._____
    traveled, in miles, by a vehicle traveling 65
    mph for $t$ hr.   A car travels 65 mph for 3 hr.
    How many miles did the car travel?

4.  The formula $A = lw$ gives the area of a                4._____
    rectangle with length $l$ and width $w$.  A
    rectangle has a length of 4 m and a
    width of $\frac{1}{2}$ m.  What is the area of the
    rectangle?

5.  The formula $I = Prt$ gives the simple interest        5._____
    earned by an investment with principal $P$,
    interest rate $r$, and time $t$.  How much simple
    interest is earned on $2000 invested at 6% for
    2 yr?  (Use 0.06 for 6%.)

**6.** The cost for one month of Will's cell phone, in dollars, is given by the formula $c = 45 + 0.1m$, where $m$ is the number of text messages sent or received that month. How much was his cell phone bill for a month in which he sent or received 80 text messages?

6._____

*Solve each formula for the indicated letter.*

**7.** $y = mx$, for $m$

7._____

**8.** $s = qrx$, for $r$

8._____

**9.** $y = 26 - x$, for $x$

9._____

**10.** $A = 4\pi r^2$, for $r^2$

10._____

**11.** $T = 5d^2$, for $d^2$          **11.**_____

**12.** $d = \dfrac{a-b}{c}$, for $b$          **12.**_____

**13.** $A = \dfrac{p+q+r}{3}$, for $q$          **13.**_____

**14.** $v = 2x + 20t$, for $t$          **14.**_____

**15.** $T = \dfrac{a}{b}$, for $a$          **15.**_____

**16.** $c = \dfrac{4k}{w}$, for $k$                              **16.** _____

**17.** $r = \dfrac{2}{3}x + 11$, for $x$                        **17.** _____

**18.** $k = ac + bc$, for $c$                            **18.** _____

**19.** $v = m + abm$, for $m$                          **19.** _____

**20.** $x = a + \dfrac{5(b-c)}{d}$, for $c$                 **20.** _____

# Chapter 2  EQUATIONS, INEQUALITIES, AND PROBLEM SOLVING

## 2.4     Applications with Percent

**Topics**
Converting Between Percent Notation and Decimal Notation
Solving Percent Problems

**Key Terms**
Use the vocabulary terms listed below to complete each statement in Exercises 1–2.

> **left**          **right**          **write**          **drop**

1.   To convert from decimal notation to percent notation, move the decimal place two places to the _____ and _____ the percent symbol.

2.   To convert from percent notation to decimal notation, move the decimal place two places to the _____ and _____ the percent symbol.

*Convert the percent notation in each sentence to decimal notation*

3.   20% of players are left-handed.                    3._____

4.   Of all registered voters, 30% voted in the election.     4._____

5.   A self-employed person must earn 17% more than a non-self-employed person performing the same task(s).                              5._____

6.   At Bella's Boutique, 16% of items sold are returned.                            6._____

*Convert to decimal notation.*
**7.** 78%

**7.**_____

**8.** 42.8%

**8.**_____

**9.** 0.55%

**9.**_____

*Convert the decimal notation in each sentence to percent notation.*
**10.** In the Johnson family, 0.18 of the children
choose tacos as their favorite food.

**10.**_____

**11.** Of all CDs purchased, 0.372 of them are
pop/rock.

**11.**_____

**12.** In Lincoln, 0.52 of the refuse is recycled.

**12.**_____

**13.** 0.91 of blood is water.

**13.**_____

*Convert to percent notation.*
**14.** 0.58

**14.**_____

**15.** $\dfrac{5}{8}$

**15.**_____

**16.** 2.00                                    **16.**_____

*Solve.*

**17.** What percent of 240 is 60?              **17.**_____

**18.** What percent of 160 is 56?              **18.**_____

**19.** 1.2 is 30% of what number?              **19.**_____

**20.** 26.25 is 17.5% of what number?          **20.**_____

**21.** What number is 65% of 2?                **21.**_____

**22.** What number is 4% of 840?               **22.**_____

**23.** 72 is what percent of 180?              **23.**_____

**24.** What is 2% of 2?                        **24.**_____

*Lightning strikes kill people in various locations. The graph details these locations and the percentage of lightning deaths that occur. From 1960 to 1992, 2897 people were killed.*

**Deaths Due to Lightning**

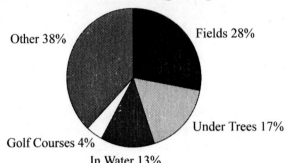

*In each of Exercises 25–27, determine the number of people killed from 1960 to 1992 in each location. Round to the nearest whole number.*

**25.** In fields

25._____

**26.** Under trees

26._____

**27.** At golf courses

27._____

**28.** Of the 8760 hours in a year, one television was on for 2628 hours. What percent is this?

28._____

**29.** Jeanne left a 12% tip on a meal. Including the tip, its cost was $50.40. What was the cost of the meal before the tip?

29._____

**30.** A low-calorie bread has 138 calories in a 3-slice serving. This is 10% fewer than the number of calories in a serving of regular bread. How many calories are in the same size serving of regular bread?

30._____

# Chapter 2  EQUATIONS, INEQUALITIES, AND PROBLEM SOLVING

## 2.5    Problem Solving

**Topics**
>  Five Steps for Problem Solving
>  Applying the Five Steps

*Solve. Even though you might find the answer quickly in some other way, practice using the five-step problem-solving process.*

1.  When 20 is subtracted from 3 times a certain number, the result is 43. What is the number?

1._____

2.  An appliance store decreases the price of a 19-in. television set 22% to a sale price of $505.44. What was the original price?

2._____

3.  A 200-m wire is cut into three pieces. The second piece is 3 times as long as the first. The third piece is 2 times as long as the second. How long is each piece?

3._____

**4.** The sum of three consecutive odd integers is 183.          **4.**_____
What are the integers?

**5.** The sum of the page numbers on the facing pages          **5.**_____
of a book is 85. What are the page numbers?

**6.** In a triangle, the second angle is 5 times as large          **6.**_____
as the first, and the third angle is 18° less than 4
times the first angle. Find the size of the angles.

**7.** The perimeter of a rectangular athletic field is 104 m and the length is 16 m more than the width. Find the length and the width.

**7.** _____

**8.** Money is borrowed at 13% simple interest. After one year, $1007.96 pays off the loan. How much was originally borrowed?

**8.** _____

**9.** A worker on the production line is paid a base salary of $240 per week plus $0.81 for each unit produced. One week the worker earned $414.15. How many units were produced?

**9.** _____

**10.** A hospital parking lot charges $2.50 for the first hour or part thereof, and $1.25 for each additional hour or part thereof. A weekly pass costs $36 and allows unlimited parking for 7 days. If each visit Charles makes to the hospital lasts one and a half hours, what is the minimum number of visits that would make it worthwhile to buy a pass?

**10.**_____

**11.** The sum of the measures of two complementary angles is 90°. If one angle measures 15° more than twice the measure of the other, find the measure of the smaller angle.

**11.**_____

**12.** The equation $R = -0.028t + 20.8$ can be used to predict the world record in the 200 meter dash, where $R$ stands for the record in seconds and $t$ stands for the number of years since 1920. According to the equation, in what year did the record become 20.24 seconds?

**12.**_____

# Chapter 2   EQUATIONS, INEQUALITIES, AND PROBLEM SOLVING

## 2.6   Solving Inequalities

**Topics**

Solutions of Inequalities
Graphs of Inequalities
Solving Inequalities Using the Addition Principle
Solving Inequalities Using the Multiplication Principle
Using the Principles Together

*Determine whether each number is a solution of the given inequality.*

**1.**  $x \le -3$                                    **1.a)**_____

   **a)** 5                                          **b)**_____

   **b)** $-6$                                       **c)**_____

   **c)** 0                                          **d)**_____

   **d)** $-2$                                       **e)**_____

   **e)** $-3$

*Graph on a number line.*

**2.**  $x \ge 3$                                    **2.**

**3.**  $s < -4$                                     **3.**

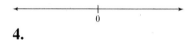

**4.**  $-2 < x \le 3$                               **4.**

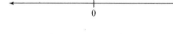

*Describe each graph using set-builder notation.*

**5.**                                               **5.**_____

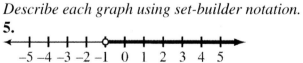

**6.**                                               **6.**_____

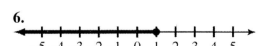

*Solve using the addition principle. Graph and write set-builder notation for each answer.*

**7.** $x + 6 > 3$

**7.**

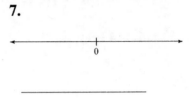

_____

**8.** $2x + 4 \leq x + 7$

**8.**

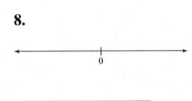

_____

*Solve using the addition principle. Write set-builder notation for each answer.*

**9.** $y + \dfrac{2}{7} \leq \dfrac{7}{14}$

**9.**

**10.** $-10z + 5 > 5 - 11z$

**10.**

_____

*Solve using the multiplication principle. Graph and write set-builder notation for each answer.*

**11.** $-12x > -36$

**11.**

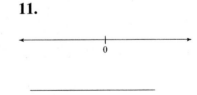

_____

**12.** $3x \geq 18$

**12.**

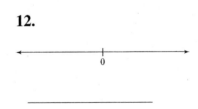

_____

*Solve using the multiplication principle. Write set-builder notation for each answer.*

**13.** $9x \geq -8$                        **13.**_____

**14.** $\dfrac{-4}{7} > -6x$                **14.**_____

*Solve using the addition and multiplication principles. Write set-builder notation for each answer.*

**15.** $6 + 9n < -21$                     **15.**_____

**16.** $8x - 8 \leq 32$                    **16.**_____

**17.** $9x - 9 < -54$                      **17.**_____

**18.** $5x + 2 - 4x \leq 13$               **18.**_____

**19.** $6 - 20x \le 5 - 11x - 8x$      **19.**_____

**20.** $\dfrac{x}{4} - 1 \le \dfrac{1}{2}$      **20.**_____

**21.** $5(2x - 3) < 45$      **21.**_____

**22.** $5(2y - 9) \ge 6(3y + 4)$      **22.**_____

# Chapter 2  EQUATIONS, INEQUALITIES, AND PROBLEM SOLVING

## 2.7    Solving Applications with Inequalities

**Topics**

Translating to Inequalities

Solving Problems

*Translate to an inequality.*

**1.** My salary next year will be at least $45,000.

**1.** _____

**2.** A number is greater than $-1.2$.

**2.** _____

**3.** The average speed, $s$ was between 80 and 120 mph.

**3.** _____

**4.** The price of a baseball glove is at most $36.49.

**4.** _____

**5.** The temperature is at most $84°$.

**5.** _____

**6.** The barrel of peaches weighs at least 12 lb.

**6.** _____

*Use an inequality and the five-step process to solve each problem.*

**7.** Susan is certain that every time she parks in the municipal garage it costs her at least $2.10. If the garage charges 30¢ plus 30¢ for each half hour, how long is Susan's car parked?

**7.** _____

**8.** A car rents for $30 per day plus 23¢ per mile. You are on a daily budget of $76. What mileage will allow you to stay within your budget?

**8.**_____

**9.** You are taking a math course in which there will be four tests, each worth 100 points. You have scores of 97, 94, and 97 on the first three tests. You must earn a total of 360 points in order to get an A. What scores on the last test will give you an A?

**9.**_____

**10.** The width of a rectangle is fixed at 6 cm. Determine those lengths for which the area will be less than 126 square cm.

**10.**_____

**11.** In planning for a banquet, you find that one speaker charges $225 plus 50% of the total ticket sales. Another speaker charges a flat fee of $525. In order for the first speaker to produce more profit for your organization than the other speaker, what is the highest price you can charge per ticket, assuming that 200 people will attend?

**11.**_____

**12.** Most insurance companies will replace a vehicle any time an estimated repair exceeds 80% of its "blue book" value. Melissa's car had $6500 in repairs after an accident. What can be concluded about its "blue book" value?

12._____

**13.** A reduced fat cookie contains 2 g of fat per serving. In order for food to be termed "reduced fat" it must have at least 25% less fat than the regular item. What can you conclude about how much fat is in a serving of the regular cookie?

13._____

**14.** The formula $I = 2(s + 10)$ can be used to convert dress sizes $s$ in the United States to dress sizes $I$ in Italy. For what dress sizes in the United States will dress sizes in Italy be larger than 50?

14._____

**15.** A phone company charges $11.71 for monthly service plus 2.2 cents per minute for local calls during peak hours, 9 A.M. to 9 P.M. The charge for off-peak local calls is 0.5 cents per minute. The maximum charge is $34.16. If only off-peak local calls are made, how long must a customer speak on the phone if the maximum charge is to apply? Round to the nearest whole minute.

15._____

**16.** Carlos can be paid in one of two ways. Plan A is a salary of $500 per month, plus a commission of 8% of sales. Plan B is a salary of $653 per month, plus a commission of 5% of sales. For what amount of sales is Carlos better off selecting Plan A?

**16.**_____

**17.** Judith is about to invest $27,000, part at 4% and the rest at 8%. What is the most she can invest at 4% and still be guaranteed at least $2000 a year?

**17.**_____

**18.** A landscaping company is laying out a triangular flower bed. The height of the triangle is 12 ft. What lengths of the base will make the area at least 168 ft$^2$?

**18.**_____

**19.** The length of a rectangle is fixed at 11 cm. What widths will make the perimeter greater than 88 cm?

**19.**_____

# Chapter 3  INTRODUCTION TO GRAPHING

## 3.1    Reading Graphs, Plotting Points, and Scaling Graphs

**Topics**

Problem Solving with Bar, Circle, and Line Graphs
Points and Ordered Pairs
Numbering the Axes Appropriately

**Key Terms**

Use the vocabulary terms listed below to complete each statement in Exercises 1–3.

origin          first          second          *x*-axis          *y*-axis

1.   The _____ is a number line that is often aligned vertically.

2.   In the pair (4, 5), the _____ coordinate is 4.

3.   The point at which the axes cross is called the _____.

*The bar graph shows the average daily expenses for lodging, food, and rental car for traveling executives.*

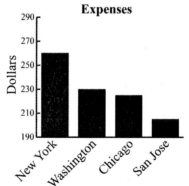

4.   Which city is the least expensive of the four?          4._____

5.   What can be concluded about an executive
     returning from a one-day trip to one of these four
     cities whose expense report is for $261.59?          5._____

                                                          _____

                                                          _____

*The pie chart shows how federal income tax dollars are used.*

**Federal Income Tax Allocation**

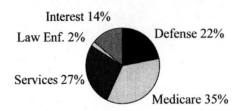

6. The O'Donnell's taxable income of $107,200 is          6._____
   taxed at an effective rate of 24%. How much of
   their earnings will go toward Medicare?

7. Sarah's taxable income of $31,400 is taxed at an        7._____
   effective rate of 16%. How much of her earnings
   will go toward defense?

*The circle graph shows music preferences of customers on the basis of music store sales.*

**Record Sales**

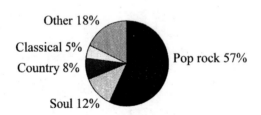

8. A music store sells 3500 recordings a month.           8._____
   How many are classical?

9. A music store sells 2500 recordings a month.           9._____
   How many are pop/rock?

*The line graph shows a company's predicted sales (in millions of dollars) for several years.*

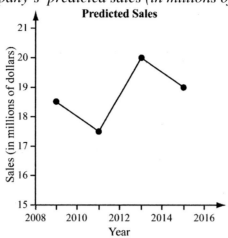

**10.** What are the predicted sales for 2011?          **10.**_____

**11.** In what year are predicted sales the greatest?          **11.**_____

*Plot the group of points.*
**12.** $(-1, 4)$, $(-5, -2)$, $(6, 3)$, $(4, -2)$, $(1, -5)$, $(0, 4)$          **12.**

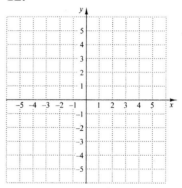

*Make a line graph of the data in the given table, listing years on the horizontal scale.*

**13.**                                                    **13.**

| Year | Tree Height (in inches) |
|------|-------------------------|
| 2001 | 74 |
| 2002 | 82 |
| 2003 | 92 |
| 2004 | 101 |
| 2005 | 113 |
| 2006 | 123 |

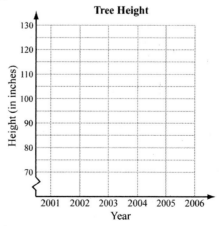

*Find the coordinates of points A, B, C, D, and E.*
**14.**

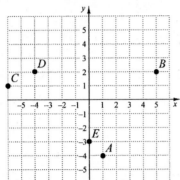

**14.**_____

_____

_____

_____

_____

*In Exercises 15–16, use a grid 10 squares wide and 10 squares high to plot the given coordinates. Choose your scale carefully. Scales may vary.*
**15.**  $(-65, 2)$, $(8, 8)$, $(-27, -5)$

**15.**

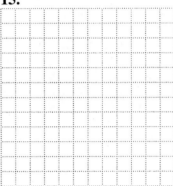

**16.**  $(4, -56)$, $(-2, 33)$, $(3, -17)$

**16.**

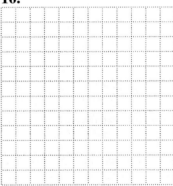

*In which quadrant is each point located?*
**17.**  $(-3, 5)$

**17.**_____

**18.**  $(1, -6)$

**18.**_____

**19.**  $(-2, -3)$

**19.**_____

# Chapter 3  INTRODUCTION TO GRAPHING

## 3.2     Graphing Linear Equations

---

**Topics**
> Solutions of Equations
> Graphing Linear Equations
> Applications

---

*Determine whether each equation has the given ordered pair as a solution.*

**1.**  $y = 10x + 13$;  $(3, 10)$                    **1.**_____

**2.**  $6p + 9q = -8$;  $(6, -8)$               **2.**_____

*In Exercises 3–5, an equation and two ordered pairs are given. Show that each pair is a solution of the equation. Then graph the two pairs to determine another solution. Answers may vary.*

**3.**  $y = 19x - 13$;  $(1, 6)$,  $(2, 25)$         **3.**_____

**4.**  $3p - 8q = -11$;  $(-9, -2)$,  $(-1, 1)$    **4.**_____

**5.**  $6p + 5q = 10$;  $(-5, 8)$,  $(0, 2)$        **5.**_____

*Graph each equation.*

**6.** $y = x + 4$

**6.**

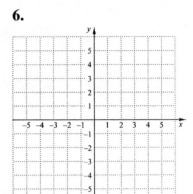

**7.** $y = \dfrac{2}{3}x + 1$

**7.**

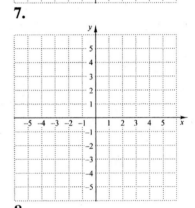

**8.** $y + x = -2$

**8.**

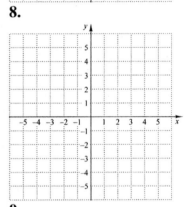

**9.** $y = x - 2$

**9.**

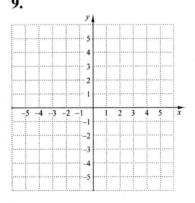

**10.** $y = 2x$

**10.**

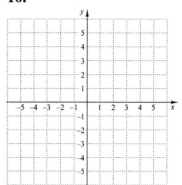

**11.** $y = \dfrac{1}{3}x - 6$

**11.**

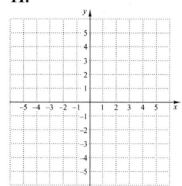

**12.** $y = 4x + 3$

**12.**

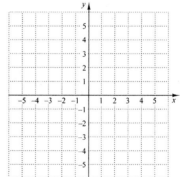

**13.** $y + x = -10$

**13.**

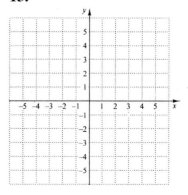

**14.** $y = 3x + 2$

**14.**

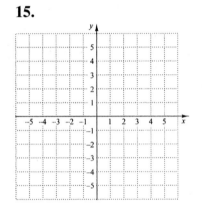

**15.** $y = \dfrac{1}{5}x - 6$

**15.**

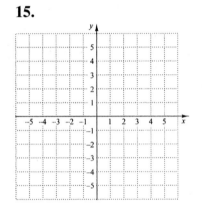

*Graph the solutions of each equation.*
**16. a)** $11 = 3x + 5$

**16. a)**

**b)** $y = 3x + 5$

**b)**

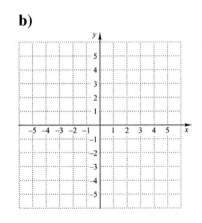

*Graph each equation using a graphing calculator. Remember to solve for y first if necessary.*

**17.** $y = -2x + 1$

**17.**

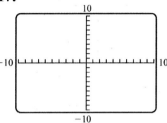

**18.** $y = 3 - x^2$

**18.**

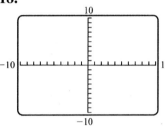

*Graph each equation using both viewing windows indicated. Determine which window best shows both the shape of the graph and where the graph crosses the x- and y-axes.*

**19.** $y = -x - 8$

    **a)** $[-10, 10, -10, 10]$, Xscl = 1, Yscl = 1

    **b)** $[-20, 20, -20, 20]$, Xscl = 5, Yscl = 5

**19.**_____

**20.** $y = 4x^2 - 13$

    **a)** $[-10, 10, -10, 10]$, Xscl = 1, Yscl = 1

    **b)** $[-20, 20, -20, 20]$, Xscl = 5, Yscl = 5

**20.**_____

*Solve by graphing. Label all axes, and show where each solution is located on the graph.*

**21.** The value *v*, in hundreds of dollars, of a shopkeeper's inventory software program is given by $v = -\frac{1}{3}t + 6$, where *t* is the number of years since the shopkeeper bought the program. What is the program worth 6 years after it is first purchased?

**21.**_____

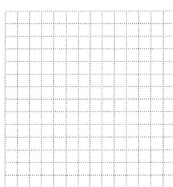

**22.** A regular smoker is 12 times more likely to die from lung cancer than a nonsmoker. An ex-smoker who stopped smoking *t* years ago is *w* times more likely to die from lung cancer than a nonsmoker, where $t + w = 12$. Cheryl gave up smoking 2 years ago. How much more likely is she to die from lung cancer than Paula, who never smoked?

**22.**_____

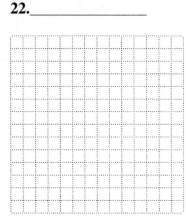

**23.** The cost, *T*, in hundreds of dollars, of tuition and fees at many community colleges can be approximated by $T = \frac{6}{5}c + 1,$ where *c* is the number of credits for which a student registers. Estimate how much tuition and fees will cost a student who registers for two three-credit courses. Round to the nearest hundred dollars.

**23.**_____

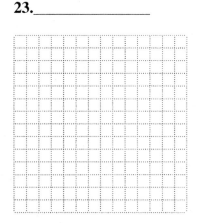

# Chapter 3 INTRODUCTION TO GRAPHING

### 3.3    Graphing and Intercepts

> **Topics**
> Intercepts
> Using Intercepts to Graph
> Graphing Horizontal or Vertical Lines

**Key Terms**
Use the vocabulary terms listed below to complete each statement in Exercises 1–3.

| horizontal line | linear equation | vertical line |

1.   A _____ is an equation whose graph is a straight line.

2.   The graph of $y = b$ is a _____.

3.   The graph of $x = a$ is a _____.

*For Exercises 4 and 5, list* **(a)** *the coordinates of the y-intercept and* **(b)** *the coordinates of all x-intercepts.*

4.

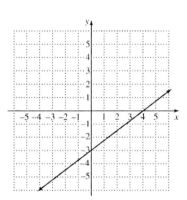

**4. a)**_____

   **b)**_____

5.

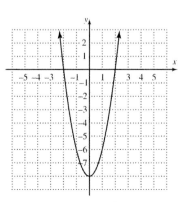

**5. a)**_____

   **b)**_____

*For Exercises 6–9, list* **(a)** *the coordinates of any y-intercept and* **(b)** *the coordinates of any x-intercept. Do not graph.*

**6.**   $6x + 5y = 30$                    **6. a)**_____

                                        **b)**_____

**7.**   $3x - 8y = 24$                    **7. a)**_____

                                        **b)**_____

**8.**   $x = 12$                          **8. a)**_____

                                        **b)**_____

**9.**   $y = -8$                          **9. a)**_____

                                        **b)**_____

*Find the intercepts.  Then graph.*

**10.**  $x - y = 2$                       **10.**_____

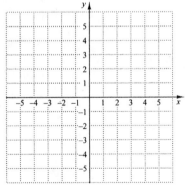

**11.**  $y = 3x - 6$

**11.** _____

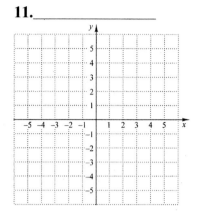

**12.**  $2x - 5y = -10$

**12.** _____

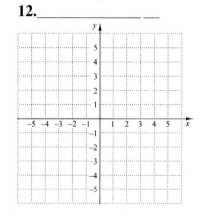

**13.**  $2x - 3y = 0$

**13.** _____

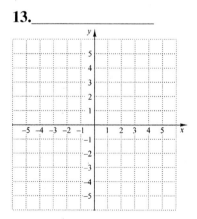

*Graph.*

**14.** $y = 4$

**14.**

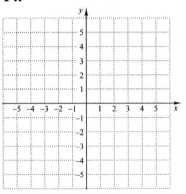

**15.** $x = 8$

**15.**

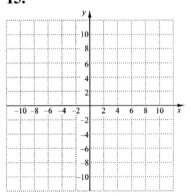

**16.** $5x = -10$

**16.**

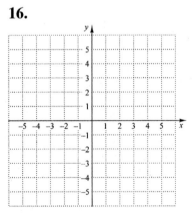

**17.** $18 + 6y = 0$

**17.**

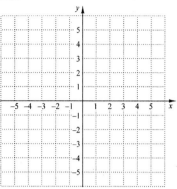

*Write an equation for each graph.*

**18.**

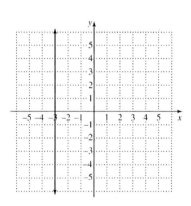

**18.**_____

**19.**

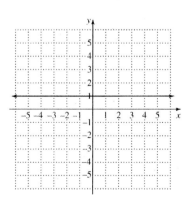

**19.**_____

*For each equation, find the x-intercept and the y-intercept. Then determine which of the given viewing windows will show both intercepts.*

**20.** $y = -15x + 50$                              **20.**_____

       **a)** $[-10,10,-10,10]$   **b)** $[-5,5,-20,80]$
       **c)** $[-60,60,-10,10]$   **d)** $[-10,0,0,100]$

**21.** $y = 0.2x + 40$                               **21.**_____

       **a)** $[-10,10,-10,10]$   **b)** $[-50,50,-200,400]$
       **c)** $[-5,5,-50,50]$     **d)** $[-500,500,-50,50]$

*Using a graphing calculator, graph each equation so that both intercepts can be easily viewed. Adjust the window settings so that tick marks can be clearly seen on both axes.*

**22.** $y = 8x - 40$                                  **22.**

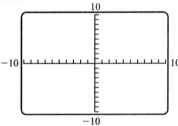

**23.** $5y - 22x = 70$                                **23.**

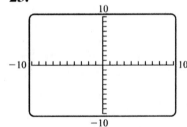

# Chapter 3 INTRODUCTION TO GRAPHING

## 3.4     Rates

| Topics |
| --- |
| Rates of Change |
| Visualizing Rates |

*Solve.*

**1.** On May 12, Charles rented a car with a full tank of gas and 12,678 mi on the odometer. On May 18, he returned the car with 12,786 mi on the odometer. The rental agency charged Charles $240 and needed 6 gal of gas to fill up the tank. Assume that the pickup time was later in the day than the return time so that no late fees were applied.

**a)** Find the car's rate of gas consumption, in miles per gallon.

**1. a)**_____

**b)**_____

**c)**_____

**b)** Find the average cost of the rental, in dollars per day.

**c)** Find the rate of travel, in miles per day

**2.** At 2:00, Hester rented a bicycle from Perry's Wheels. She returned the bike at 5:00, after riding 33 mi. Hester paid $13.50 for the rental.

**a)** Find Hester's average speed, in miles per hour.

**2. a)**_____

**b)**_____

**c)**_____

**b)** Find the rental rate, in dollars per hour.

**c)** Find the rental rate, in dollars per mile.

**3.** Carlos reports to work at 7:00 A.M. and leaves at
6:00 P.M. after typing from the end of page 61 to
the end of page 127. His employer pays $121.00
for his work.
**a)** Find the rate of pay, in dollars per hour.

**3. a)**_____

**b)**_____

**c)**_____

**b)** Find the average typing rate, in number of
pages per hour.

**c)** Find the rate of pay, in dollars per page.

**4.** The tuition at a local community college was
$2472 in 1993 and $3423 in 1997. Find the rate at
which tuition was increasing.

**4.**_____

**5.** At 1:23, Lisa exits the lobby at the corner of
Moneo Drive and 6$^{th}$ Street and steps into a
waiting taxicab. At 1:32, she leaves the cab and
enters a hotel at the corner of Moneo and 51$^{st}$ St.
**a)** Find the cab's average rate of travel, in streets
per minute.

**5. a)**_____

**b)**_____

**b)** Find the cab's rate of travel, in seconds per
street.

**6.** Mark and Walt hiked Crown Mountain, elevation 4033 ft, from the West trailhead, elevation 1223 ft. They began at 8:00 A.M. and reached the summit at 1:59 P.M.

a) Find their average rate of ascent, in feet per minute, rounded to the nearest hundredth.

**6. a)** _____

**b)** _____

b) Find their average rate of ascent, in minutes per foot, rounded to the nearest hundredth.

*In Exercises 7–10, draw a linear graph to represent the given information. Be sure to label and number the axes appropriately.*

**7.** At 7:00 P.M., the Chicago to London express flew out over the Atlantic Ocean at 600 miles per hour.

**7.**

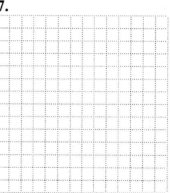

**8.** In 1994, there were 640 crimes in Otis County, a figure which was dropping at a rate of about 40 per year.

**8.**

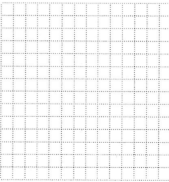

**9.** By 1:00 P.M., Francine had earned $90. She continued earning money at the rate of $30 per hour.

**9.**

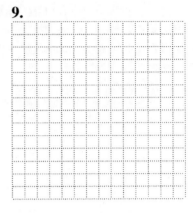

**10.** Mike's phone bill was already $23.00 when he made a call for which he was charged at a rate of $0.25 per minute.

**10.**

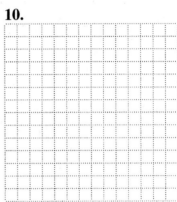

*In Exercises 11–15, use the graph provided to calculate a rate of change in which the units of the horizontal axis are used in the denominator.*

**11.** The graph shows data for a car driven in a city. Find the rate of gas consumption for the car in gallons per mile.

**11.**_____

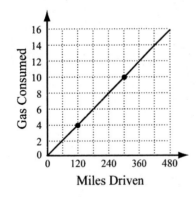

**12.** The graph shows data regarding the value of a certain computer over time. What is the rate of decrease of the value of the computer in dollars per year?

12._____

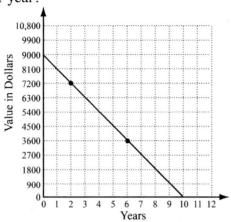

**13.** The graph shows data from a recent satellite internet connection. Find the billing rate in cents per minute.

13._____

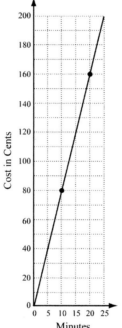

**14.** The manager at a cycle repair shop has a graph displaying data from a recent day of work. That day, approximately how many bikes did the shop repair per hour?

14._____

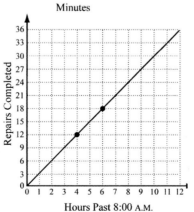

**15.** The graph shows data from an afternoon on the site of a construction job. The electrical team installed cable at a rate of how many feet per hour?

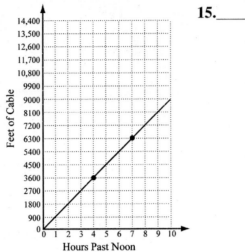

**15.** _____

*Match each description with the most appropriate graph. Scales are intentionally omitted.*

**(a)**

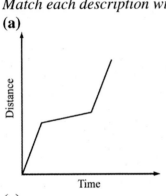

**(b)**

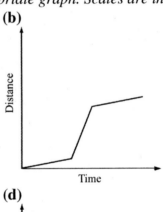

**(c)**

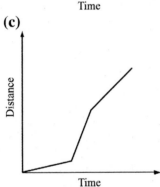

**(d)**

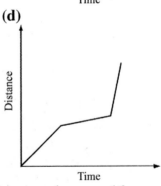

**16.** On Saturday, Ron walked to a friend's house, rode a bike from there to the store and back, then ran home.

**16.** _____

**17.** On Sunday, Ron ran to the store, then walked to a friend's house, then rode a bike home.

**17.** _____

**18.** On Monday, Ron walked to school, then ran to the library, then walked home.

**18.** _____

**19.** On Tuesday, Ron rode his bike to school, walked to the store and back to school, then rode his bike home.

**19.** _____

# Chapter 3  INTRODUCTION TO GRAPHING

## 3.5     Slope

> **Topics**
>    Rate and Slope
>    Horizontal and Vertical Lines
>    Applications

**1.** Find the rate of change of a country's population.          **1.**

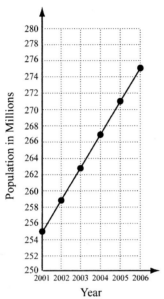

**2.** Find the rate of change of a country's defense outlays.          **2.**

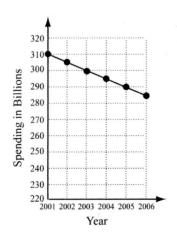

**3.** Find the rate of change of the tuition and fees at private four-year colleges.

**3.**_____

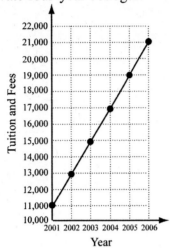

*Find the slope, if it is defined, of each line. If the slope is undefined, state this.*

**4.**

**4.**_____

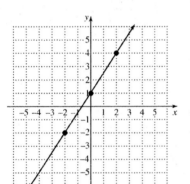

**5.**

**5.**_____

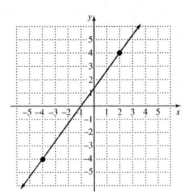

**6.**

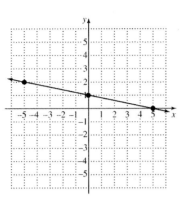

**6.**_____

**7.**

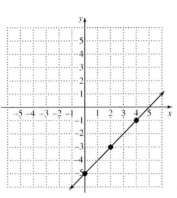

**7.**_____

**8.**

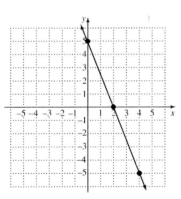

**8.**_____

**9.**

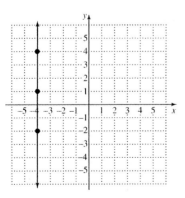

**9.**_____

**10.**

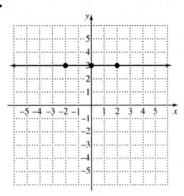

**10.**_____

**11.**

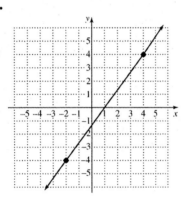

**11.**_____

**12.**

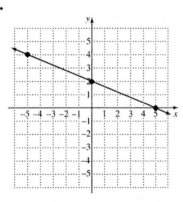

**12.**_____

*Find the slope of the line containing each given pair of points. If the slope is undefined, state this.*

**13.** $(7,0)$ and $(8,10)$

**13.**_____

**14.** $(9,0)$ and $(10,6)$

**14.**_____

**15.** $(14,-19)$ and $(-18,-20)$

**15.**_____

**16.** $(10,5)$ and $(4,0)$

**16.**_____

**17.** $(11,-15)$ and $(-11,-17)$

**17.**_____

**18.** $\left(-2,\dfrac{1}{2}\right)$ and $\left(-8,\dfrac{1}{2}\right)$

**18.**_____

**19.** $(5,-8)$ and $(5,-9)$

**19.**_____

*In Exercises 20–22, find the slope of each line whose equation is given. If the slope is undefined, state this.*

**20.** $x = -7$                         **20.**_____

**21.** $y = 10$                         **21.**_____

**22.** $x = 6$                          **22.**_____

**23.** A mountain path drops 346.8 ft vertically for    **23.**_____
every 5509 ft horizontally. What is the grade of
the path?

**24.** A ramp rises 1.3 ft vertically and 6.5 ft        **24.**_____
horizontally. Express the slope as a grade.

**25.** A roof rises 1.6 ft vertically and 11.3 ft       **25.**_____
horizontally. What is the grade of the roof?

**26.** A mountain peak in Colorado rises from sea level  **26.**_____
to a summit elevation of 14,368 ft over a
horizontal distance of 15,835 ft. Find the grade of
the peak.

# Chapter 3 INTRODUCTION TO GRAPHING

## 3.6    Slope–Intercept Form

**Topics**

> Using the *y*-intercept and the Slope to Graph a Line
> Equations in Slope–Intercept Form
> Graphing and Slope–Intercept Form

**Key Terms**

Use the vocabulary terms listed below to complete each statement in Exercises 1–4.

| | |
|---|---|
| **parallel** | **perpendicular** |
| **slope** | ***y*-intercept** |

1. The _____ of a line written in the form $y = mx + b$ is given by *m*.

2. The _____ of a line written in the form $y = mx + b$ is given by *b*.

3. Two lines with different *y*-intercepts are _____ if they have the same slope.

4. Two lines are _____ if the product of their slopes is $-1$.

*Draw a line that has the given slope and y-intercept.*

5. Slope: $\dfrac{3}{4}$; *y*-intercept $(0, 2)$

**5.**

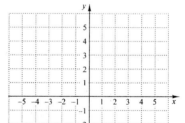

6. Slope: $\dfrac{5}{2}$; *y*-intercept $(0, -3)$

**6.**

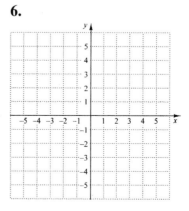

**7.** Slope: $3$; $y$-intercept $(0, -5)$

**7.**

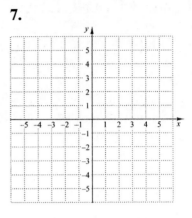

**8.** Slope: $-2$; $y$-intercept $(0, 3)$

**8.**

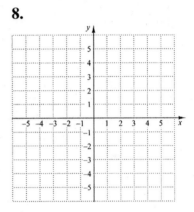

*Find the slope and the y-intercept of each line whose equation is given.*

**9.** $y = -\dfrac{3}{5}x + 2$

**9.** _____

**10.** $y = \dfrac{4}{9}x - 5$

**10.** _____

**11.** $4x - 5y = 9$

**11.** _____

**12.** $2x + 3y = 9$

**12.** _____

*Find the slope–intercept equation for the line with the indicated slope and y-intercept.*

**13.** Slope $\dfrac{2}{9}$; *y*-intercept $(0,6)$

13._____

**14.** Slope $-3$; *y*-intercept $(0,-4)$

14._____

**15.** Slope $-\dfrac{3}{5}$; *y*-intercept $(0,-\frac{1}{4})$

15._____

**16.** Slope $5$; *y*-intercept $(0,\frac{2}{5})$

16._____

*Determine an equation for each graph shown.*

**17.**

17._____

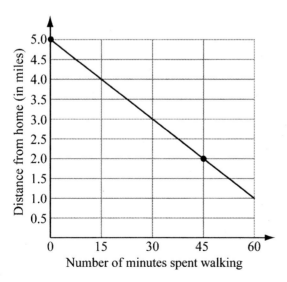

**18.**

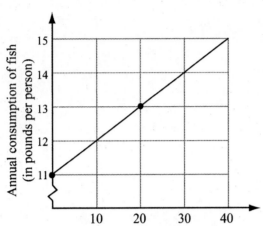

Number of years since 1960
Source: Based on data from the
National Marine Fisheries Service

**18.**

*Graph.*

**19.** $y = \dfrac{3}{2}x - 5$

**19.**

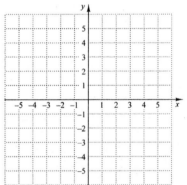

**20.** $y = \dfrac{2}{3}x - 1$

**20.**

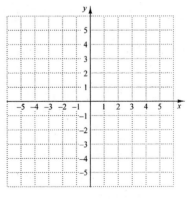

**21.** $y = -2x + 3$

**21.**

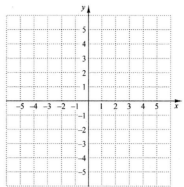

**22.** $y = -\dfrac{3}{4}x$

**22.**

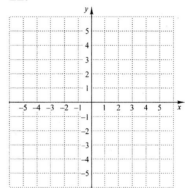

**23.** $3x - y = 4$

**23.**

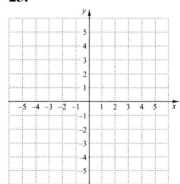

**24.** $4x + 3y = 6$

**24.**

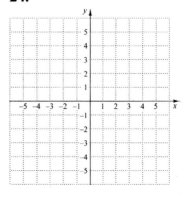

*Determine whether each pair of equations represents parallel lines.*

**25.**  $y = 3 - 2x$,
$y - 2x = 5$

**25.**_____

**26.**  $x - y = -2$,
$5 + x = y$

**26.**_____

*Determine whether each pair of equations represents perpendicular lines.*

**27.**  $y = 2x - 1$,
$x + 2y = 3$

**27.**_____

**28.**  $x + 3y = 3$,
$2x = 6y - 1$

**28.**_____

*Write a slope–intercept equation of the line whose graph is described.*

**29.** Parallel to the graph of $y = -2x + 1$;
$y$-intercept $(0, -3)$

**29.**_____

**30.** Parallel to the graph of $2x - 4y = 5$;
$y$-intercept $(0, 1)$

**30.**_____

**31.** Perpendicular to the graph of $y = 3x - 5$;
$y$-intercept $(0, 2)$

**31.**_____

**32.** Perpendicular to the graph of $3y + 2x = 7$;
$y$-intercept $(0, -4)$

**32.**_____

# Chapter 3 INTRODUCTION TO GRAPHING

### 3.7    Point–Slope Form

---

**Topics**

Writing Equations in Point–Slope Form
Graphing and Point–Slope Form
Estimations and Predictions Using Two Points

---

**Key Terms**

Use the terms or expressions listed below to complete each statement in Exercises 1–5.

| $m$ | $(x_1, y_1)$ | extrapolation |
|---|---|---|
| **interpolation** | **point-slope form** | |

1.  An equation of the form $y - y_1 = m(x - x_1)$ is written in _____.

2.  The slope of the graph of an equation of the form $y - y_1 = m(x - x_1)$ is

    _____.

3.  The point _____ is on the graph of an equation of the form
    $y - y_1 = m(x - x_1)$.

4.  Estimating a value between known points is called _____.

5.  Estimating a value beyond known points is called _____.

*Write a point–slope equation for the line with the given slope that contains the given point.*

6.  $m = 3, (5, 1)$                          6._____

7.  $m = -2, (3, -7)$                        7._____

8.  $m = -\frac{3}{4}, (-6, 2)$              8._____

9.  $m = \frac{2}{5}, (-4, -8)$              9._____

*For each point–slope equation listed, state the slope and a point on the graph.*

**10.**  $y - 7 = \frac{1}{5}(x - 2)$                              **10.**_____

**11.**  $y + 3 = -3(x - 1)$                                   **11.**_____

**12.**  $y + 5 = 6(x + 1)$                                    **12.**_____

**13.**  $y = -\frac{2}{3}x$                                    **13.**_____

*Write the slope–intercept equation for the line with the given slope that contains the given point.*

**14.**  $m = 2, (-1, 1)$                                      **14.**_____

**15.**  $m = -\frac{3}{2}, (-2, -3)$                           **15.**_____

*Write the slope-intercept equation of the line that contains the specified point and is parallel to the indicated line.*

**16.**  $(2, -1); \ 4x - y = 8$                               **16.**_____

**17.**  $(-12, 0); \ 2x + 3y = 5$                             **17.**_____

*Write the slope–intercept equation for the line containing the given pair of points.*

**18.** $(3,6)$ and $(-1,-3)$                    **18.**_____

**19.** $(-4,2)$ and $(-5,-8)$                   **19.**_____

**20.** Graph the line with slope $\frac{1}{3}$ that passes through the point $(1,2)$.

**20.**

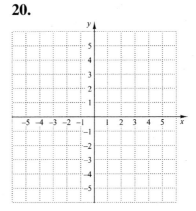

**21.** Graph the line with slope $-\frac{4}{3}$ that passes through the point $(-1,5)$.

**21.**

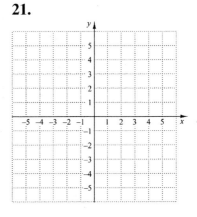

*Graph.*

**22.** $y - 1 = 2(x + 3)$

**22.**

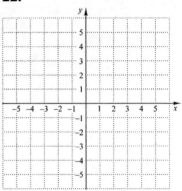

**23.** $y + 6 = -\frac{1}{3}(x - 4)$

**23.**

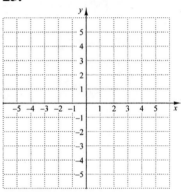

# Chapter 4 POLYNOMIALS

## 4.1     Exponents and Their Properties

| Topics |
| --- |
| Multiplying Powers with Like Bases |
| Dividing Powers with Like Bases |
| Zero as an Exponent |
| Raising a Power to a Power |
| Raising a Product or a Quotient to a Power |

*Simplify. Assume that no denominator is zero and $0^0$ is not considered.*

**1.** $u^4 \cdot u^9$                                    1._____

**2.** $3^3 \cdot 3^0$                                    2._____

**3.** $(5s)^9 \cdot (5s)^7$                              3._____

**4.** $\left(a^7 b^8\right)\left(a^5 b^6\right)$         4._____

**5.** $(x+3)^5 (x+3)^2$                                  5._____

**6.** $x^4 \left(xy^2\right)(xy)$                        6._____

**7.** $\dfrac{u^8}{u^4}$

7._____

**8.** $\dfrac{6^9 r^5}{6^5 r^3}$

8._____

**9.** $\dfrac{3^5 s^5}{3^2 s^3}$

9._____

**10.** $\dfrac{20m^6}{5m^3}$

10._____

**11.** $\dfrac{x^8 y^7}{x^2 y^4}$

11._____

**12.** $\dfrac{w^9 z^8}{w^2 z^4}$

12._____

*Simplify.*

**13.** $s^0$ when $s = -6$

**13.**_____

**14.** $3m^0$ when $m = -3$

**14.**_____

**15.** $12^1 - 12^0$

**15.**_____

*Simplify. Assume that no denominator is zero and $0^0$ is not considered.*

**16.** $\left(m^{28}\right)^6$

**16.**_____

**17.** $(2c)^3$

**17.**_____

**18.** $(-2a)^9$

**18.**_____

**19.** $\left(5q^9\right)^4$

**19.**_____

**20.** $\left(8c^{10}d^{14}\right)^3$

**20.**_____

**21.** $\left(3x^{15}y^{20}\right)^4$

**21.** _____

**22.** $\left(\dfrac{c^4}{2}\right)^2$

**22.** _____

**23.** $\left(\dfrac{3}{7m}\right)^3$

**23.** _____

**24.** $\left(\dfrac{x^6}{y^7}\right)^4$

**24.** _____

**25.** $\left(\dfrac{n^2}{8}\right)^2$

**25.** _____

**26.** $\left(\dfrac{n^5a}{r}\right)^9$

**26.** _____

**27.** $\left(\dfrac{a^4}{-3b^3}\right)^4$

**27.** _____

# Chapter 4 POLYNOMIALS

## 4.2     Polynomials

**Topics**
> Terms
> Types of Polynomials
> Degree and Coefficients
> Combining Like Terms
> Evaluating Polynomials and Applications

**Key Terms**
Use the vocabulary terms listed below to complete each statement in Exercises 1–4.

  **term     coefficient     degree     polynomial**

1.   A _____ is a monomial or a sum of monomials.

2.   A polynomial composed of two _____(s) is a binomial.

3.   The _____ of a term is the number of variable factors in that term.

4.   The constant factor of a term is the _____ of that term.

5. Determine whether the expression is a           5._____
   polynomial:   $5x^{-2} - 6.9$

6. Identify the terms of the polynomial:           6._____
   $3 - 4x + x^2$

7. Determine the coefficient and the degree of each   7._____
   term in the polynomial:   $5x^2 + 6x + 3$

8. For the polynomial $x^5 - 9x + x^9 - 7x^7$,      8. a)_____
   **a)** List the degree of each term.
   **b)** Determine the leading term and the leading    b)_____
        coefficient.
   **c)** Determine the degree of the polynomial.        c)_____

*Classify each polynomial as a monomial, binomial, trinomial, or none of these.*

**9.**  $x^2 - 10x + 25$

**9.**_____

**10.**  $64x^6 - 49$

**10.**_____

*Combine like terms. Write all answers in descending order.*

**11.**  $5x^5 - 5x^6 - 4x^5 + 6x^6$

**11.**_____

**12.**  $-7x + 6x^2 - 2x + 4x^2 + 3$

**12.**_____

**13.**  $3x + 2x + 3x - x^5 - 8x^5$

**13.**_____

**14.**  $-x + \dfrac{3}{4} + 27x^7 - x - \dfrac{1}{2} - 2x^7$

**14.**_____

*Evaluate each polynomial for x = 2 and for x = −2.*

**15.**  $3x^2 - 4x + 6$

**15.**_____

**16.**  $x^2 - 8x + 1$

**16.**_____

**17.** $3x^2 - 3x + 9$ 

**18.** For a club consisting of $n$ people, the number of ways in which a president, vice president, and treasurer can be elected can be determined using $n^3 - 3n^2 + 2n$. Find the number if the club has 17 members.

**18.**_____

**19.** The distance $s$, in feet, traveled by a body falling freely from rest in $t$ seconds is approximated by the polynomial $s = 16t^2$. A brick is dropped from the top of a building and takes 1 second to hit the ground. How tall is the building?

**19.**_____

**20.** In a psychology experiment, participants were able to memorize an average of $M$ words in $t$ minutes, where $M = -0.0011t^3 + 0.07t^2$. Use the graph of $M$ to estimate the number of words memorized after 12 minutes, rounded to the nearest integer.

**20.**_____

**21.** The polynomial $M(t) = 0.5t^4 + 3.45t^3 - 96.65t^2 + 347.7t$   **21.**_____

$0 \le t \le 6$, estimates the number of milligrams of ibuprofen in the bloodstream $t$ hours after 400 mg of medication has been swallowed. Use the graph to estimate the number of milligrams of ibuprofen in the bloodstream 4 hours after 400 mg has been swallowed. Round to the nearest 40 mg.

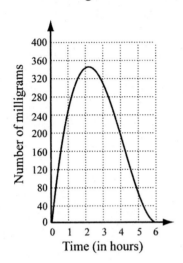

**22.** A 12-oz beverage can has a height of 7.4 inches.   **22.**_____

Evaluate the polynomial $2\pi rh + 2\pi r^2$ for $h = 7.4$ in. and $r = 1.7$ in. to find the surface area of the can. Use 3.14 as an approximation for $\pi$ and round to the nearest tenth.

# Chapter 4 POLYNOMIALS

## 4.3    Addition and Subtraction of Polynomials

| **Topics** |
| --- |
| Addition of Polynomials |
| Opposites of Polynomials |
| Subtraction of Polynomials |
| Problem Solving |

*Add.*

**1.** $(7x+5)+(-8x+2)$

1._____

**2.** $(-4x+8)+(x^2+x-9)$

2._____

**3.** $(1.2x^3+4.2x^2-4.5x)+(-3.1x^3-4.5x^2+62)$

3._____

**4.** $(7+3x+4x^2+4x^3)+(6-3x+4x^2-4x^3)$

4._____

**5.** $\left(\dfrac{1}{12}x^4+\dfrac{1}{4}x^3+\dfrac{3}{8}x^2+5\right)+\left(-\dfrac{7}{12}x^4+\dfrac{5}{8}x^2-5\right)$

5._____

**6.** $(0.05x^5-0.8x^2+x+0.08)+(-0.02x^5+x^3-0.3x-0.07)$

6._____

7. $\begin{array}{r} -6z^8 + 8z^7 + 8z - 4 \\ -4z^7 + 3z + 4 \\ \hline \end{array}$

7._____

8. Write two equivalent expressions for the
opposite of the polynomial:     $-x^2 + 16x - 3$

8._____

_____

*Simplify*

9. $-\left(4x^2 - 4x + 7\right)$

9._____

10. $-\left(2x^4 + 4x^2 + \dfrac{1}{4}x - 4\right)$

10._____

*Subtract.*

11. $(-6x + 9) - (x^2 + x - 8)$

11._____

12. $(7x^4 + 7x^3 - 2) - (9x^2 - 2x + 5)$

12._____

13. $(1.4x^3 + 4.2x^2 - 3.3x) - (-4.4x^3 - 4.2x^2 + 92)$

13._____

14. $(4n^3 + 7n) - (-9n^3 - 6n + 2)$

14._____

**15.** $\left(\dfrac{1}{6}x^3 - \dfrac{3}{8}x - \dfrac{1}{6}\right) - \left(-\dfrac{1}{6}x^3 + \dfrac{3}{8}x - \dfrac{1}{6}\right)$     **15.**_____

**16.** $(0.08x^3 - 0.05x^2 + 0.05x) - (0.07x^3 + 0.08x^2 - 5)$     **16.**_____

**17.**
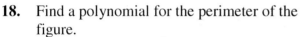
$$x^2 + 6x + 5$$
$$-(x^2 + 2x)$$

**17.**_____

**18.** Find a polynomial for the perimeter of the figure.     **18.**_____

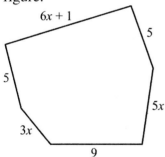

**19.** Find two algebraic expressions for the area of the figure. First, regard the figure as one large rectangle, and then regard the figure as a sum of four smaller rectangles.     **19.**_____

_____

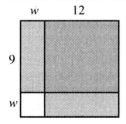

**20.** A 2-ft by 3-ft bath mat is placed on the floor of a bathroom measuring $x$ ft by $x$ ft. Find a polynomial expression for the floor area not covered by the mat.

20._____

**21.** A 6-ft-wide circular garden is planted in a lawn that is $w$ ft by $w$ ft. Find a polynomial for the area of the lawn outside the garden.

21._____

# Chapter 4 POLYNOMIALS

## 4.4    Multiplication of Polynomials

**Topics**
   Multiplying Monomials
   Multiplying a Monomial and a Polynomial
   Multiplying Any Two Polynomials
   Checking by Evaluating

*Multiply.*

**1.** $(2x^9)(6)$

1. _____

**2.** $(-x^9)(-x)$

2. _____

**3.** $(0.4x^2)(0.4x^8)$

3. _____

**4.** $\left(-\dfrac{1}{4}x^4\right)\left(-\dfrac{1}{3}x\right)$

4. _____

**5.** $(-6x^7)(0)$

5. _____

**6.** $(8x^7)(-4x^7)(7x^6)$

6. _____

**7.** $2x(-x+7)$

**8.** $(x+6)5x$

**9.** $x^3(x^5+1)$

**10.** $4x(8x^2-2x+6)$

**11.** $(-4x^5)(x^5+x)$

**12.** $\dfrac{2}{3}a^3\left(12a^6-3a^4-\dfrac{5}{7}\right)$

**13.** $(x+4)(x+2)$

**14.** $(x+6)(x-3)$

**15.**  $(x-1)(x-8)$                    **15.**_____

**16.**  $(x+19)(x-19)$                  **16.**_____

**17.**  $\left(\dfrac{3}{7}a+4\right)\left(\dfrac{4}{7}a-3\right)$   **17.**_____

*Draw and label rectangles to illustrate each product.*
**18.**  $x(x+4)$

**19.**  $(x+6)(x+3)$

*Multiply and check.*

**20.** $\left(x^2+x+7\right)\left(x-7\right)$

**20.**_____

**21.** $\left(4x+2\right)\left(2x^2+3x+1\right)$

**21.**_____

**22.** $\left(y^2-8\right)\left(4y^2-9y+2\right)$

**22.**_____

**23.** $\left(-8x^3-5x^2+2\right)\left(5x^2-x\right)$

**23.**_____

**24.** $\left(3+x+x^2\right)\left(-3-x+x^2\right)$

**24.**_____

**25.** $\left(6t^2-t-6\right)\left(9t^2+6t-1\right)$

**25.**_____

**26.** $\left(x^3+x^2+x+1\right)\left(8x-8\right)$

**26.**_____

# Chapter 4 POLYNOMIALS

## 4.5    Special Products

---
**Topics**

     Products of Two Binomials
     Multiplying Sums and Differences of Two Terms
     Squaring Binomials
     Multiplications of Various Types

---

*Multiply.*

**1.** $\left(x^8 + 4\right)(x + 9)$

1._____

**2.** $(y + 4)(y - 5)$

2._____

**3.** $(6x + 4)(6x + 4)$

3._____

**4.** $(4t - 5)(4t + 5)$

4._____

**5.** $(4x - 6)(x - 9)$

5._____

**6.** $\left(p-\dfrac{1}{5}\right)\left(p+\dfrac{1}{5}\right)$

6._____

**7.** $(-8x+7)(x+7)$

7._____

**8.** $\left(x^2+9\right)\left(x^3-8\right)$

8._____

**9.** $\left(5x^6+7\right)\left(9x^2+5\right)$

9._____

*Multiply. Try to recognize what type of product each multiplication is before multiplying.*

**10.** $(x+7)(x-7)$

10._____

**11.** $\left(6x^9+3\right)\left(6x^9-3\right)$

11._____

**12.** $\left(x^5 - x^2\right)\left(x^5 + x^2\right)$

12._____

**13.** $(x+8)^2$

13._____

**14.** $\left(2x^2 + 2\right)^2$

14._____

**15.** $\left(a - \dfrac{9}{2}\right)^2$

15._____

**16.** $5n^4\left(7n^2 - 1\right)$

16._____

**17.** $\left(x^2 - 4\right)\left(x^2 + x - 1\right)$

17._____

**18.** $\left(5-5x^8\right)^2$

**19.** $9x\left(x^2+7x-5\right)$

**20.** $\left(5x^7+3\right)\left(x+9\right)$

**21.** $\left(9-5x^3\right)^2$

**22.** $\left(y-7\right)\left(y^2+7y+49\right)$

**23.** $\left(7x^7+6\right)^2$

**24.** $\left(y-5\right)\left(y^2+5y+25\right)$

*Find the total area of all shaded rectangles*

**25.**

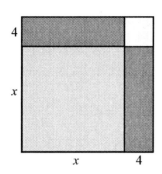

**25.**_____

**26.**

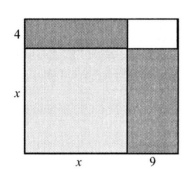

**26.**_____

**27.**

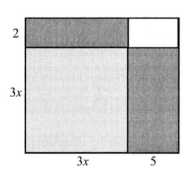

**27.**_____

*Draw and label rectangles similar to those in Exercises 25–27 to illustrate each of the following.*

**28.** $(x+6)^2$

**29.** $(11+t)^2$

# Chapter 4 POLYNOMIALS

## 4.6  Polynomials in Several Variables

---

**Topics**
    Evaluating Polynomials
    Like Terms and Degree
    Addition and Subtraction
    Multiplication

---

**1.** Evaluate the polynomial for $x = 3$ and $y = -2$:         **1.**_____
$$x^2 + y^2 + xy$$

**2.** Evaluate the polynomial for $x = 4$, $y = -6$, and     **2.**_____
$z = -2$:    $xyz^2 + z$

*The polynomial $0.041h - 0.018a - 2.69$ can be used to estimate the lung capacity, in liters, of a female with height h, in centimeters, and age a, in years.*

**3.** Find the lung capacity of a 50-year-old woman     **3.**_____
who is 170 cm tall.

**4.** Find the lung capacity of a 30-year-old woman     **4.**_____
who is 150 cm tall.

**5.** A launched rocket has an altitude, in meters, given by the polynomial $h + vt - 4.9t^2$, where $h$ is the height, in meters, from which the launch occurs, at a velocity $v$ in meters per second, and $t$ is the number of seconds for which the rocket is airborne. If a rocket is launched from the top of a tower 130 m high with an initial upward speed of 30m/sec, what will its height be after 2 sec?

**5.** _____

*Identify the coefficient and the degree of each term of each polynomial. Then find the degree of each polynomial.*

**6.** $x^7 y - 8xy + 3x^3 - 1$

**6.** _____

_____

**7.** $22x^4 y^8 - 6x^4 yz - 9$

**7.** _____

_____

*Combine like terms.*

**8.** $a + b - 6a - 8b$

**8.** _____

**9.** $3x^7 y - 2xy^7 + x^3$

**9.** _____

**10.** $6u^6 v - 5uv^7 + 5u^6 v - 6uv^7$

**10.** _____

*Add or subtract, as indicated.*

**11.** $\left(8x^2 - xy + y^2\right) + \left(-x^2 - 6xy + 2y^2\right)$          **11.**_____

**12.** $\left(7r^2 + 7rf - 2f^2\right) - \left(9r^2 - 5rf + 11f^2\right)$          **12.**_____

**13.** $\left(6x^2 - 5xy + y^2\right) + \left(-9x^2 - 2xy - y^2\right) + \left(x^2 + xy - 9y^2\right)$     **13.**_____

*Multiply.*

**14.** $(5z - u)(3z + 5u)$          **14.**_____

**15.** $\left(a^5 b - 6\right)\left(a^5 b - 7\right)$          **15.**_____

**16.** $\left(2m^2 - 8n^2\right)(9m + 2n)$          **16.**_____

**17.** $(5x+6h)^2$

**18.** $\left(r^9t^7-9\right)^2$

**19.** $\left(v^2+tj\right)\left(v^2-tj\right)$

**20.** $(x+y-12)(x+y+12)$

**21.** $[x+y+1][x-(y+1)]$

**22.** $\left(-bc+a^2\right)\left(bc+a^2\right)$

*Find the total area of each shaded region.*

**23.**

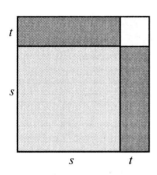

**23.**_____

**24.**

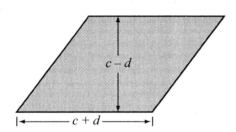

**24.**_____

**25.**

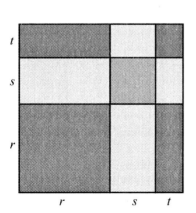

**25.**_____

*Draw and label rectangles similar to the rectangle in Exercise 23 to illustrate each of the following.*

**26.** $(a+b)(c+d)$

**27.** $(x+y)(x+z)$

# Chapter 4  POLYNOMIALS

## 4.7    Division of Polynomials

---

**Topics**
>    Dividing by a Monomial
>    Dividing by a Binomial

---

**Key Terms**
Use the vocabulary terms listed below in Exercises 1–4 to label each part of the division shown.

|  dividend | divisor | quotient | remainder |
|-----------|---------|----------|-----------|

1. _____

2. _____

3. _____

4. _____

$$\begin{array}{r}
x + 2 \longleftarrow ① \\
x + 3 \overline{\smash{\big)}\, x^2 + 5x + 10} \longleftarrow ③ \\
\underline{x^2 + 3x} \\
2x + 10 \\
\underline{2x + \phantom{0}6} \\
4 \longleftarrow ④
\end{array}$$

② $\longrightarrow x + 3$

*Divide and check.*

5.  $\dfrac{15a^8 - 40a^3}{5}$

5. _____

6.  $\dfrac{7t - t^5 + 3t^6}{t}$

6. _____

7.  $(6x^5 - 30x^4 + 12x) \div (3x)$

7. _____

**8.**  $(15y^5 - 5y^3 - 30y) \div (-5y^2)$

**8.**_____

**9.**  $\dfrac{24x^7 - 40x^5 - 18x^3}{4x^3}$

**9.**_____

**10.**  $\dfrac{36s^5t^2 - 60s^7t^3 + 42s^4t^4}{-6s^4t^2}$

**10.**_____

**11.**  $(x^2 - 2x - 80) \div (x - 10)$

**11.**_____

**12.**  $(p^2 - 4p + 12) \div (p - 3)$                    **12.**_____

**13.**  $(a^3 - 2a + 3) \div (a + 1)$                    **13.**_____

**14.**  $(x^3 - 4x^2 + 5) \div (x - 2)$                    **14.**_____

**15.** $(15x^2 + 2x - 8) \div (5x + 4)$

**16.** $(12x^4 + 8x^3 - 10x^2 + 5) \div (2x - 1)$

**17.** $(2x^4 + 3x^3 - 5x^2 - x - 6) \div (x^2 - 3)$

**18.** $(5x^3 - 5x^2 - 20x + 2x^4 - 12) \div (x^2 - 4)$

**18.**＿＿＿＿＿＿＿＿＿＿

**19.** $(x^3 - 3x^2 + x - 2) \div (x - 2)$

**19.**＿＿＿＿＿＿＿＿＿＿

**20.** $(y^2 + 5y + 4) \div (y + 4)$

**20.**＿＿＿＿＿＿＿＿＿＿

**21.** $(a^3 - 5a^2 + 4a - 3) \div (a + 3)$

**21.**_____

**22.** $(2x^3 + 5x - 1) \div (x - 3)$

**22.**_____

**23.** $(t^4 - 16) \div (t - 2)$

**23.**_____

**24.** $(4x^3 + 1 + x - 2x^2) \div (x + \frac{1}{2})$

**24.**_____

# Chapter 4 POLYNOMIALS

## 4.8    Negative Exponents and Scientific Notation

**Topics**
> Negative Integers as Exponents
> Scientific Notation
> Multiplying and Dividing Using Scientific Notation

*Express using positive exponents. If possible, simplify.*

**1.** $\left(-\dfrac{3}{4}\right)^{-4}$

1._____

**2.** $x^3 y^{-2}$

2._____

**3.** $\dfrac{1}{r^{-t}}$

3._____

**4.** $\left(\dfrac{2}{5}\right)^{-3}$

4._____

*Express using negative exponents.*

**5.** $\dfrac{1}{y^2}$

5._____

**6.** $\dfrac{1}{b^p}$

6._____

**7.** $\dfrac{1}{5}$

7._____

*Simplify. Do not use negative exponents in the answer.*

**8.** $5^{-3} \cdot 5^9$

8._____

**9.** $s^{-6} \cdot s^{-7}$

9._____

**10.** $\left(4y^{-6}\right)^2$

10._____

**11.** $\dfrac{x^5}{x^{-8}}$

11._____

**12.** $\left(\dfrac{y^6 x}{za^2}\right)^{-3}$

12._____

**13.**  $\left(\dfrac{s^3 p}{tw^5}\right)^{-6}$

13._____

**14.**  $\left(x^{-2}z^{-3}\right)^{-3}$

14._____

**15.**  $\left(\dfrac{25x^9 y^{-7}}{5x^{-2} y^{-2}}\right)^0$

15._____

**16.**  Convert to decimal notation:  $8.18 \times 10^7$

16._____

**17.**  Convert to scientific notation:  0.048000

17._____

*Multiply or divide and write scientific notation for the result.*

**18.**  $\left(2 \times 10^5\right)\left(3 \times 10^5\right)$

18._____

**19.**  $\left(1.5 \times 10^3\right)\left(8.7 \times 10^{-5}\right)$

19._____

**20.**  $\dfrac{9.2\times10^{-3}}{2.3\times10^{17}}$                    **20.**_____

# Chapter 5 POLYNOMIALS AND FACTORING

## 5.1    Introduction to Factoring

**Topics**
>    Factoring Monomials
>    Factoring When Terms Have a Common Factor
>    Factoring by Grouping
>    Checking by Evaluating

**Key Terms**
Use the vocabulary terms listed below to complete each statement in Exercises 1–3.

| factor | factorization | largest common factor |
|--------|---------------|----------------------|

1.   To _____ a polynomial is to write it as a product.

2.   When a polynomial is written as a product, that product is called a(n) _____ of the polynomial.

3.   When factoring a polynomial, we first factor out the _____.

*Find three factorizations for each monomial. Answers may vary.*

4.    $18x^3$                          4._____

     _____

     _____

5.    $-21a^{10}$                      5._____

     _____

     _____

6.    $72x^8$                          6._____

     _____

     _____

*Factor. Remember to use the largest common factor and to check by multiplying.*
*Factor out a negative factor if the first coefficient is negative.*

**7.**   $4t^3 - 20t^2$

**7.**_____

**8.**   $6x^2y - 10xy^2$

**8.**_____

**9.**   $7a^2 - 14a + 49$

**9.**_____

**10.**   $15x^5yz^5 - 10x^2y^4z^6 + 25x^3y^3z^4$

**10.**_____

**11.**   $-10x + 40$

**11.**_____

**12.**   $-6x^2 + 36x + 48$

**12.**_____

**13.**   $3a - 3b$

**13.**_____

**14.**   $-t^4 - t^3 + 2t + 13$

**14.**_____

**15.** $20m^4 - 24m^3 - 28m^2$                    **15.**_____

**16.** $-5x^4 + 40x^3$                            **16.**_____

**17.** $7x^7 - 49x^5$                             **17.**_____

**18.** $3x^6 + 3x - 12$                           **18.**_____

*Factor.*

**19.** $x(y+2) + z(y+2)$                          **19.**_____

**20.** $(p-2)(p+4) + (p-2)(p+6)$                  **20.**_____

**21.** $6t^3(t-1) + 5(1-t)$                       **21.**_____

*Factor by grouping, if possible, and check.*

**22.** $uv + vw + tu + tw$                        **22.**_____

**23.** $x^3 - 5x^2 - 2x + 10$

**23.** _____

**24.** $a^5 - a^4 - 5a^2 + 5a^3$

**24.** _____

**25.** $5pq - p^2q + 4p - 20$

**25.** _____

# Chapter 5 POLYNOMIALS AND FACTORING

## 5.2    Factoring Trinomials of the Type $x^2 + bx + c$

**Topics**

When the Constant Term Is Positive
When the Constant Term Is Negative
Prime Polynomials

**Key Terms**

Use the vocabulary terms listed below to complete each statement in Exercises 1–6.

| | | |
|---|---|---|
| **binomial** | **common factor** | **constant term** |
| **factorization** | **prime** | **trinomial** |

1.    A polynomial with three terms is a _____.

2.    A polynomial with two terms is a _____.

3.    A _____ polynomial cannot be factored.

4.    When factoring, always check first for a _____.

5.    In the polynomial $x^2 + 5x + 6$, the _____ is 6.

6.    The _____ of $x^2 + 5x + 6$ is $(x+2)(x+3)$.

*Factor completely. Remember to look first for a common factor. Check by multiplying. If a polynomial is prime, state this.*

7.    $x^2 + 11x + 18$                           7._____

8.    $t^2 - 13t + 40$                            8._____

**9.** $x^3 - 4x^2 - 12x$

**9.** _____

**10.** $5n^2 + 25n - 70$

**10.** _____

**11.** $44 + a^2 - 15a$

**11.** _____

**12.** $t^2 + 8t + 11$

**12.** _____

**13.** $15x + 36 + x^2$

**13.** _____

**14.** $42 - n - n^2$

**14.** _____

**15.** $3x - 18 + x^2$

**15.** _____

**16.** $12z^3 - 24z^2 - 180z$

**16.** _____

**17.** $a^2 - 9ab - 22b^2$

**17.** _____

**18.** $y^5 - 74y^4 + 73y^3$

**18.** _____

**19.** $w^2 + 8w + 15$

**19.** _____

**20.** $r^2 + 7r + 6$

**20.** _____

**21.** $s^2 - 12s + 32$

**21.** _____

**22.** $c^2 + 3c - 28$

**22.** _____

**23.** $3s^2 - 30s + 63$

**23.** _____

**24.** $2a + a^2 - 35$

**24.** _____

**25.** $-b^3 + b^2 + 72b$

**25.** _____

# Chapter 5 POLYNOMIALS AND FACTORING

### 5.3    Factoring Trinomials of the Type $ax^2 + bx + c$

| Topics |
|---|
| Factoring with FOIL |
| The Grouping Method |

*Factor completely. If a polynomial is prime, state this.*

**1.**  $10x^2 + x - 3$                                   **1.**_____

**2.**  $6a^3 - 11a^2 - 10a$                              **2.**_____

**3.**  $40t^2 - 38t - 15$                                 **3.**_____

**4.**  $5x - 15 + 20x^2$                                  **4.**_____

**5.**  $25w^2 + 19w + 4$                                  **5.**_____

**6.**  $4x^2 + 16x + 15$                                  **6.**_____

**7.** $-18z^2 + 24z + 10$

**7.** _____

**8.** $12 - 20x - 25x^2$

**8.** _____

**9.** $8a^3b - 2a^2b - 15ab$

**9.** _____

**10.** $22x^2 - 4 + 87x$

**10.** _____

**11.** $30t^4 - 63t^3 - 30t^2$

**11.** _____

**12.** $6p^2 - 19pq + 15q^2$

**12.** _____

**13.** $25c^2 - 20cd + 4d^2$

**13.** _____

**14.** $8a^2b^2 + 13ab - 6$

*Factor. Use factoring by grouping even though it would seem reasonable to first combine like terms.*

**15.** $v^2 + 5v + 6v + 30$

**15.**_____

**16.** $4r^2 - 7r - 28r + 49$

**16.**_____

**17.** $6s^2 + 4s + 9s + 6$

**17.**_____

**18.** $21r^2 - 6r + 14r - 4$

**18.**_____

**19.** $30b^2 - 35b + 18b - 21$

**19.**_____

*Factor completely. If a polynomial is prime, state this.*

**20.** $35m^5 + 9m^4 - 2m^3$

20._____

**21.** $9m^2 + 6m - 24$

21._____

**22.** $900b + 80b^2 - 20b^3$

22._____

**23.** $28b^2 + 111b - 4$

23._____

**24.** $27a^5 - 93a^4 - 22a^3$

24._____

**25.** $56c^2 - 23cv - 63v^2$

25._____

# Chapter 5 POLYNOMIALS AND FACTORING

**5.4    Factoring Perfect-Square Trinomials and Differences of Squares**

| Topics |
| --- |
| Recognizing Perfect-Square Trinomials |
| Factoring Perfect-Square Trinomials |
| Recognizing Differences of Squares |
| Factoring Differences of Squares |
| Factoring Completely |

**Key Terms**

Use the vocabulary terms listed below to complete each statement in Exercises 1–3.

**difference of squares          factored completely          perfect-square trinomial**

1.  An expression that does not contain any factors that can be factored further is

    _____.

2.  The expression $x^2 - 36$ is an example of a _____.

3.  The expression $x^2 - 12x + 36$ is an example of a _____.

*Determine whether each of the following is a perfect-square trinomial.*

4.  $w^2 - 6w + 9$                                    **4.** _____

5.  $w^2 + 6w - 9$                                    **5.** _____

6.  $a^2 - 8a + 64$                                    **6.** _____

7.  $16b^2 - 400b + 500$                          **7.** _____

*Factor completely. Remember to look first for a common factor and to check by multiplying. If a polynomial is prime, state this.*

**8.** $x^2 - 2x + 1$                      **8.**_____

**9.** $y^2 + 16 + 8y$                 **9.**_____

**10.** $2a^2 + 12a + 18$              **10.**_____

**11.** $25 - 10n + n^2$               **11.**_____

**12.** $y^2 + 4xy + 4x^2$              **12.**_____

**13.** $100x + x^3 + 20x^2$           **13.**_____

**14.** $0.01c^2 - 0.10c + 0.25$       **14.**_____

*Determine whether each of the following is a difference of squares.*

**15.** $v^2 - 49$                         **15.**_____

**16.** $a^2 + 16$

**17.** $b^2 - 32$

*Factor completely. Remember to look first for a common factor. If a polynomial is prime, state this.*

**18.** $t^2 - 1$

**18.**＿＿＿＿＿＿＿＿＿

**19.** $9 - m^2 n^2$

**19.**＿＿＿＿＿＿＿＿＿

**20.** $5ab^2 - 20a$

**20.**＿＿＿＿＿＿＿＿＿

**21.** $15x^4 - 15y^4$

**21.**＿＿＿＿＿＿＿＿＿

**22.** $64x^2 y^8 - 25x^6$

**22.**＿＿＿＿＿＿＿＿＿

**23.** $\dfrac{1}{4} - y^2$

**23.**＿＿＿＿＿＿＿＿＿

**24.**   $(y-z)^2 - 64$                          **24.** _____

*Factor completely. If a polynomial is prime, state this.*

**25.**   $x^3 + x^2 - 9x - 9$                     **25.** _____

**26.**   $p^2 + 10pq + 25q^2 - 100t^2$            **26.** _____

**27.**   $8s^2 - 98$                              **27.** _____

**28.**   $-8s^2 + s^3 + 16s$                      **28.** _____

**29.**   $9s^2 - 16f^2$                           **29.** _____

**30.**   $a^4 - 2401$                             **30.** _____

# Chapter 5 POLYNOMIALS AND FACTORING

## 5.5    Factoring: A General Strategy

**Topics**
>   Choosing the Right Method

**Key Terms**
Use the vocabulary terms listed below to complete each statement in Exercises 1–4.

| | |
|---|---|
| **common factor** | **difference of squares** |
| **factor completely** | **grouping** |

1.   When factoring a polynomial, always look first for a(n) _____.

2.   If there are two terms in a polynomial, try first to factor as a(n) _____.

3.   If there are four terms in a polynomial, try to factor by _____.

4.   Always _____.

*Factor completely. If a polynomial is prime, state this.*

5.   $2x^2 - x - 6$                                  5._____

6.   $3y^2 - 12$                                     6._____

7.   $a^3 - 12a^2 + 36a$                             7._____

8.   $x^2 + 81 + 18x$                                8._____

**9.** $x^3 - 2x^2 - 9x + 18$

**10.** $8x^3 - 50x$

**11.** $n^2 + 100$

**12.** $6t^3 + 26t^2 - 20t$

**13.** $12x^2 - 80x - 28$

**14.** $-3y^5 + 24y^4 - 48y^3$

**15.** $c^4 - 25$

**16.** $3t^6 - 48t^2$

**17.** $c^2 x + c^2 y$

**18.** $14p^2q - 7pq$

**19.** $2x(5x - y) - (5x - y)$

**20.** $r^2 + 5r + rs + 5s$

**21.** $t^2 - 4t + 4 - 100v^2$

**22.** $2x^2 + xy - 15y^2$

**23.** $9p^2 - 12pq + 4q^2$

**24.** $a^2 + 12ab + 36b^2$

**25.** $x^2 - 3x + 6$

**26.** $2x^3 - 6x^2y + 5xy^2$

**27.** $20m^2 - 7mn - 6n^2$                    **27.** _____

**28.** $-x^2 - 5x + 24$                        **28.** _____

**29.** $a^4 - b^6 - 10b^3 - 25$                **29.** _____

**30.** $9t^2 - 8t + \frac{16}{9}$              **30.** _____

**31.** $\frac{1}{9}a^2 + \frac{4}{15}a + \frac{4}{25}$   **31.** _____

**32.** $x^4 + 5x^3 + 5x^2 + 25x$               **32.** _____

# Chapter 5 POLYNOMIALS AND FACTORING

## 5.6     Solving Quadratic Equations by Factoring

**Topics**
> The Principle of Zero Products
> Factoring to Solve Equations

*Solve using the principle of zero products.*

**1.**    $(x-1)(x-9)=0$                   **1.**_____

**2.**    $t(t-5)(t+7)=0$              **2.**_____

**3.**    $3x(x-9)=0$                   **3.**_____

*Solve by factoring and using the principle of zero products.*

**4.**    $x^2-6x=16$                   **4.**_____

**5.**    $n^3+n^2-6n=0$              **5.**_____

**6.**  $a^2 = 12a$

**6.**_____

**7.**  $50 - x^2 = 23x$

**7.**_____

**8.**  $x^2 + 12x = 0$

**8.**_____

**9.**  $-12x^2 = 24x$

**9.**_____

**10.**  $6x^4 = 8x^5$

**10.**_____

**11.**  $3x^2 + 2x - 5 = 0$

**11.**_____

**12.** $3x^3 - 11x^2 - 4x = 0$            **12.** _____

**13.** $16x^2 + 34x = 15$            **13.** _____

**14.** $2x(4x + 5) = 7$            **14.** _____

**15.** $x^2 + 36 = 12x$            **15.** _____

**16.** $x^2 = \dfrac{1}{49}$            **16.** _____

**17.** $x^2 - 64 = 0$            **17.** _____

**18.**  $x^3 + 3x^2 = 4x + 12$

**18.**_____

**19.**  Use this graph to solve $-x^2 - 2x + 3 = 0$.

**19.**_____

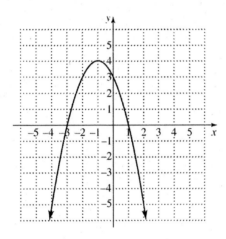

**20.**  Use this graph to solve $x^2 + 4x - 5 = 0$.

**20.**_____

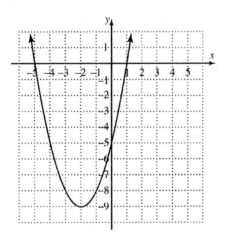

# Chapter 5 POLYNOMIALS AND FACTORING

### 5.7    Solving Applications

| **Topics** |
| --- |
| Applications |
| The Pythagorean Theorem |

**Key Terms**
Use the vocabulary terms listed below to complete each statement in Exercises 1–4.

| hypotenuse | leg | Pythagorean theorem | right triangle |
| --- | --- | --- | --- |

1. A(n) _____ contains a 90° angle.

2. The _____ of a right triangle is the side opposite the right angle.

3. A _____ of a right triangle forms one of the sides of the right angle.

4. The _____ states that the sum of the squares of the lengths of the legs of a right triangle is equal to the square of the length of the hypotenuse.

*Solve.*

5. The square of a number plus the number is 240. What is the number?

5._____

6. A scrapbook is 3 in. longer than it is wide. Find the length and the width if the area is 108 $in^2$.

6._____

**7.** A picture frame measures 16 cm by 20 cm, and 192 cm$^2$ of picture shows. Find the width of the frame.

7._____

**8.** A rectangular playground is 100 ft by 70 ft. Part of the playground is torn up to plant trees in a strip of uniform width around the playground. The area of the new playground is 4000 ft$^2$. How wide is the strip of trees?

8._____

**9.** Three consecutive odd integers are such that the square of the first plus the square of the third is 170. Find the three integers.

9._____

**10.** A wire is stretched from the ground to the top of a pole. The wire is 34 ft long. The height of the pole is 14 ft greater than the distance $d$ from the pole's base to the bottom of the wire. Find the distance $d$ and the height of the pole.

10._____

**11.** A triangular sign is 2 ft taller than it is wide. The area is 24 ft$^2$. Find the height and the base of the sign.

**11.**_____

**12.** A rectangular yard is 20 ft longer than it is wide. Determine the dimensions of the yard if it measures 100 ft diagonally.

**12.**_____

**13.** The foot of an extension ladder is 15 ft from a wall. The ladder is 5 ft longer than the height that it reaches on the wall. How far up the wall does the ladder reach?

**13.**_____

**14.** A flower bed is to be 8 m longer than it is wide. The flower bed will have an area of 84 m$^2$. What will its dimensions be?

**14.**_____

**15.** Bright Ideas determines that the revenue $R$, in dollars, from the sale of $x$ lamps is given by $2x^2 - x$. If the cost $C$, in dollars, of producing $x$ lamps is given by $x^2 + x + 8$, how many lamps must be produced and sold in order for the company to break even?

**15.** _____

**16.** The height $h(t)$, in feet, of a flare launched upward with an initial velocity of 48 ft/sec from a height of 160 ft after $t$ seconds can be approximated by $h(t) = -16t^2 + 48t + 160$. After how long will the flare reach the ground?

**16.** _____

**17.** The number of barrels of U.S. ethanol-fuel production, in millions, $t$ years after 1997, can be approximated by

$B(t) = t^2 - 3t + 35$

(*Source*: Based on data from the Energy Information Administration). How many years after 1997 were 45 million barrels of ethanol fuel produced?

**17.** _____

# Chapter 6  RATIONAL EXPRESSIONS AND EQUATIONS

## 6.1     Rational Expressions

---
**Topics**
> Simplifying Rational Expressions
> Factors that are Opposites
---

**Key Terms**
Use the vocabulary terms listed below to complete each statement in Exercises 1–2.

> **rational expression**          **simplify**          **cancel**

1.  To _____ a rational expression, factor and remove factors equal to 1.

2.  A polynomial divided by a nonzero polynomial is called a _____.

*List all numbers for which each rational expression is undefined.*

3.  $\dfrac{3}{9w+5}$                                           3._____

4.  $\dfrac{y^2+17}{y^2-3y-40}$                                 4._____

5.  $\dfrac{p^3-8p}{p^2-16}$                                    5._____

6.  $\dfrac{x^2+1}{x^2-2x-8}$                                   6._____

*Simplify, if possible. Then check by evaluating.*

**7.**  $\dfrac{24r^{10}}{32r^{12}}$                                  **7.** _____

**8.**  $\dfrac{40q^{7}}{45q^{9}}$                                   **8.** _____

**9.**  $\dfrac{5a-10}{5}$                                     **9.** _____

**10.**  $\dfrac{4y-28}{4y+28}$                               **10.** _____

**11.**  $\dfrac{5q^{2}+5q}{35q^{2}+25q}$                       **11.** _____

**12.**  $\dfrac{2x+14}{x^{2}+7x}$                             **12.** _____

**13.** $\dfrac{4t-12}{t^2-9}$

**13.**_____

**14.** $\dfrac{r^2-81}{r^2-18r+81}$

**14.**_____

**15.** $\dfrac{6n^2+6n}{42n^2+36n}$

**15.**_____

**16.** $\dfrac{14w^2-56}{49w^2-196}$

**16.**_____

**17.** $\dfrac{9-y}{y-9}$

**17.**_____

**18.** $\dfrac{t^2-49}{t^2-14t+49}$

**18.**_____

**19.** $\dfrac{x^2 - 25}{x - 5}$

**20.** $\dfrac{3t - 1}{1 - 9t^2}$

**21.** $\dfrac{5z - 15}{6z - 18}$

**22.** $\dfrac{s^2 - 1}{s^2 - 2s + 1}$

**23.** $\dfrac{x^2 - 7x + 6}{x^2 + x - 2}$

# Chapter 6  RATIONAL EXPRESSIONS AND EQUATIONS

## 6.2    Multiplication and Division

**Topics**
>    Multiplication
>    Division

**Key Terms**
Use the vocabulary terms listed below to complete each statement in Exercises 1 and 2.

**invert**                          **reciprocal**

1.  To divide two rational expressions, multiply the first expression by the
    _____ of the divisor.

2.  When we divide rational expressions, we often say that we _____ and
    multiply.

*Multiply. Leave each answer in factored form.*

3.  $\dfrac{2a}{7} \cdot \dfrac{a+1}{5a-3}$                    3._____


4.  $\dfrac{x+5}{4} \cdot \dfrac{3x+5}{x-7}$                    4._____


*Multiply and, if possible, simplify.*

5.  $\dfrac{3x^6}{5x} \cdot \dfrac{10}{x^{10}}$                    5._____


6.  $\dfrac{4t}{6t+9} \cdot \dfrac{8t+12}{t^3}$                    6._____

**7.** $\dfrac{a+b}{a-b} \cdot \dfrac{8a-8b}{a^2-b^2}$

**7.**_____

**8.** $\dfrac{x^2+1}{x^2-5x+6} \cdot \dfrac{x+1}{x-1}$

**8.**_____

**9.** $\dfrac{t^2-8t-9}{3t} \cdot \dfrac{12t^2}{t-9}$

**9.**_____

**10.** $\dfrac{x^2-2x}{x^2+5x+6} \cdot \dfrac{x^2+6x+9}{x^4-4x^2}$

**10.**_____

**11.** $\dfrac{5-y}{30y-10} \cdot \dfrac{5-45y^2}{y^2-6y+5}$

**11.**_____

**12.** $\dfrac{a^4-a}{4a-8} \cdot \dfrac{4a^2-16}{2a^4+2a^3+2a^2}$

**12.**_____

*Find the reciprocal of each expression.*

**13.** $\dfrac{10}{11x}$

**13.**_____

**14.** $t^2 - 25$

**14.**_____

*Divide and, if possible, simplify.*

**15.** $\dfrac{3}{x} \div \dfrac{x}{8}$

**15.**_____

**16.** $\dfrac{36a^8}{35a^2} \div \dfrac{45a^3}{14a}$

**16.**_____

**17.** $(x-3) \div \dfrac{x^2 - 10x + 21}{3x^2 - 3}$

**17.**_____

**18.** $\dfrac{b^2 - a^2}{8ab + 8a^2} \div (2b^2 - 3ab + a^2)$

**18.**_____

**19.** $\dfrac{x^2+4x+4}{3x-9} \div \dfrac{x^2-3x-10}{x^2-9}$

**20.** $\dfrac{x^2-9}{x^2+3x-4} \div \dfrac{x^2-7x+12}{x^2+9x+20}$

**21.** $\dfrac{x^3-8}{x^7-4x^5} \div \dfrac{x^2+2x+4}{x^2+4x+4}$

**22.** $\dfrac{27-y^3}{9-y^2} \div \dfrac{9+3y+y^2}{9-6y+y^2}$

# Chapter 6 RATIONAL EXPRESSIONS AND EQUATIONS

## 6.3    Addition, Subtraction, and Least Common Denominators

---

**Topics**
>    Addition When Denominators Are the Same
>    Subtraction When Denominators Are the Same
>    Least Common Multiples and Denominators

---

*Perform the indicated operation. Simplify, if possible.*

**1.**  $\dfrac{3}{z} + \dfrac{7}{z}$

1._____

**2.**  $\dfrac{6}{6+v} + \dfrac{7}{6+v}$

2._____

**3.**  $\dfrac{9d+89}{8d} - \dfrac{d+1}{8d}$

3._____

**4.**  $\dfrac{3z+4}{z-1} - \dfrac{z-1}{z-1}$

4._____

**5.** $\dfrac{y^2}{y-10} + \dfrac{y-110}{y-10}$

**5.** _____

**6.** $\dfrac{a^2}{a-8} - \dfrac{19a-88}{a-8}$

**6.** _____

**7.** $\dfrac{t^2+3t}{t-2} + \dfrac{2t-14}{t-2}$

**7.** _____

**8.** $\dfrac{x+2}{x^2+5x+4} + \dfrac{2}{x^2+5x+4}$

**8.** _____

**9.** $\dfrac{a^2+5}{a^2+3a-4} - \dfrac{6}{a^2+3a-4}$

**9.** _____

**10.** $\dfrac{b^2 - 8b}{b^2 + 16b + 64} + \dfrac{7b - 72}{b^2 + 16b + 64}$          **10.**_____

**11.** $\dfrac{2x^2 + 4}{x^2 - 10x + 9} - \dfrac{x^2 - 17x + 22}{x^2 - 10x + 9}$          **11.**_____

**12.** $\dfrac{6w - 13}{w^2 - 10w + 21} + \dfrac{6 - 5w}{w^2 - 10w + 21}$          **12.**_____

**13.** $\dfrac{4z - 1}{z^2 - 5z - 36} - \dfrac{z - 2}{z^2 - 5z - 36}$          **13.**_____

*Find the LCM.*
**14.** 63, 147          **14.**_____

**15.** 24, 36

**15.**_____

**16.** 27, 10, 45

**16.**_____

**17.** $10x^8$, $50x^5$

**17.**_____

**18.** $175y^5x^9$, $25y^5x^9$

**18.**_____

**19.** $5(z-11)$, $55(z-11)$

**19.**_____

**20.**  $x^2 - 36, \ x^2 + 10x + 24$

**20.**_____

**21.**  $x^9 + 4x^7, \ x^3 - 4x^2 + 4x$

**21.**_____

**22.**  $a + 6, \ (a-6)^2, \ a^2 - 36$

**22.**_____

**23.**  $x^2 + 14x + 49, \ x^2 + 4x - 21$

**23.**_____

**24.**  $14x^2 + 28x, \ 7x^2 + 49x + 70$

**24.**_____

*Find equivalent expressions that have the LCD.*

**25.** $\dfrac{22}{10x^6}, \dfrac{y}{20x^4}$

**25.** _____

**26.** $\dfrac{2}{3a^3w}, \dfrac{5}{27aw^6}$

**26.** _____

**27.** $\dfrac{t+3}{t^2-16}, \dfrac{t-4}{t^2+9t+20}$

**27.** _____

# Chapter 6  RATIONAL EXPRESSIONS AND EQUATIONS

**6.4     Addition and Subtraction with Unlike Denominators**

---

**Topics**
>     Adding and Subtracting with LCDs
>     When Factors Are Opposites

---

*Perform the indicated operation. Simplify, if possible.*

**1.**  $\dfrac{9}{v} + \dfrac{4}{v^2}$

1._____

**2.**  $\dfrac{10}{21r} - \dfrac{5}{49r}$

2._____

**3.**  $\dfrac{8}{yt^2} + \dfrac{5}{y^2 t}$

3._____

**4.**  $\dfrac{5}{36r^7} - \dfrac{4}{30r^6}$

4._____

**5.** $\dfrac{t+7}{8} + \dfrac{t-1}{28}$

**5.** _____

**6.** $\dfrac{3a-1}{2a^4} + \dfrac{5a+1}{4a^3}$

**6.** _____

**7.** $\dfrac{w-9}{w} - \dfrac{8w-67}{11w}$

**7.** _____

**8.** $\dfrac{x+s}{xs^2} + \dfrac{2x+s}{x^2s}$

**8.** _____

**9.** $\dfrac{6}{v+4} - \dfrac{4}{v-4}$

**9.** _____

**10.** $\dfrac{9}{z+1}+\dfrac{5}{3z}$                                     **10.** _____

**11.** $\dfrac{7}{x^2-9}-\dfrac{6}{x+3}$                              **11.** _____

**12.** $\dfrac{t}{t-3}-\dfrac{3}{4t-12}$                             **12.** _____

**13.** $\dfrac{3}{v-8}+\dfrac{7}{(v-8)^2}$                           **13.** _____

**14.** $\dfrac{4h}{2h-4}+\dfrac{5h}{4h-8}$                           **14.** _____

**15.** $\dfrac{w}{w^2+11w+30}-\dfrac{5}{w^2+9w+20}$

**15.** _____

**16.** $\dfrac{4x+3}{x-y}+\dfrac{3x^2-3xy}{x^2-2xy+y^2}$

**16.** _____

**17.** $\dfrac{4}{x^2+12x+36}+\dfrac{3}{x^2-36}$

**17.** _____

**18.** $\dfrac{p-5z}{p^3-z^3}-\dfrac{4z}{z^3-p^3}$

**18.** _____

**19.** $\dfrac{4-p}{9-p^2}+\dfrac{p+4}{p-3}$

**19.** _____

**20.** $\dfrac{3x+2}{5x+20} + \dfrac{x}{16-x^2}$

**20.**_____

**21.** $\dfrac{4-a^2}{a^2-16} - \dfrac{a-2}{4-a}$

**21.**_____

**22.** $\dfrac{x^2+6}{x^4-16} + \dfrac{10}{16-x^4}$

**22.**_____

**23.** $\dfrac{8-z}{z-5} - \dfrac{5z-9}{5-z}$

**23.**_____

*Perform the indicated operations. Simplify, if possible.*

**24.** $\dfrac{12y}{y^2-36} - \dfrac{6}{y} - \dfrac{6}{y+6}$

**24.**_____

**25.** $\dfrac{7w}{w^2-1}+\dfrac{2w}{1-w}-\dfrac{7}{w-1}$

**25.**_____

# Chapter 6  RATIONAL EXPRESSIONS AND EQUATIONS

## 6.5     Complex Rational Expressions

**Topics**
>     Using Division to Simplify
>     Multiplying by the LCD

*Simplify. Use either method or the method specified by your instructor.*

**1.** $\dfrac{\dfrac{x-2}{x+3}}{\dfrac{x+1}{x-5}}$

**1.** _____

**2.** $\dfrac{\dfrac{2}{p}-\dfrac{1}{q}}{\dfrac{3}{p}+\dfrac{5}{q}}$

**2.** _____

**3.** $\dfrac{\dfrac{y^2-z^2}{y}}{\dfrac{y+z}{yz}}$

**3.** _____

**4.** $\dfrac{a-\dfrac{9}{2a}}{\dfrac{3}{4a}-1}$

**4.** _____

**5.** $\dfrac{\dfrac{x^2 - y^2}{xy}}{x^{-1} - y^{-1}}$

**5.** _____

**6.** $\dfrac{\dfrac{2}{x-h} - \dfrac{2}{x}}{h}$

**6.** _____

**7.** $\dfrac{\dfrac{x^2 - x - 2}{x^2 + 8x + 15}}{\dfrac{x^2 - 4}{x^2 + 6x + 9}}$

**7.** _____

**8.** $\dfrac{\dfrac{2}{x^2 + 7x + 10} + \dfrac{x}{x^2 + 7x + 10}}{\dfrac{x}{x^2 + x - 6} - \dfrac{2}{x^2 + x - 6}}$

**8.** _____

**9.** $\dfrac{\dfrac{1}{x}-6}{11+\dfrac{1}{x}}$

**9.** _____

**10.** $\dfrac{a^{-2}+a}{a-a^{-2}}$

**10.** _____

**11.** $\dfrac{\dfrac{x}{(x-2)}-\dfrac{3}{(x+1)}}{\dfrac{x}{(x-2)}-\dfrac{1}{(x+1)}}$

**11.** _____

**12.** $\dfrac{\dfrac{5}{x^2-4}+\dfrac{3}{x-2}}{\dfrac{1}{x^2-4}+\dfrac{3}{x+2}}$

**12.** _____

**13.** $\dfrac{\dfrac{x^2}{x^2-4}+\dfrac{25}{4-x^2}}{\dfrac{x^2}{x^2-4}+\dfrac{5x}{4-x^2}}$

**13.** _____

**14.** $\dfrac{\dfrac{t}{t+3}+\dfrac{1}{5t}}{\dfrac{t}{2t+6}+\dfrac{1}{t}}$

**14.** _____

**15.** $\dfrac{\dfrac{5}{x^2-3x+2}-\dfrac{4}{x^2-5x+6}}{\dfrac{4}{x^2+x-12}-\dfrac{5}{x^2+3x-4}}$

**15.** _____

**16.** $\dfrac{\dfrac{2}{a^2+9a+20}-\dfrac{1}{a^2+a-6}}{\dfrac{2}{a^2+7a+12}-\dfrac{1}{a^2+3a-10}}$

**16.** _____

# Chapter 6 RATIONAL EXPRESSIONS AND EQUATIONS

### 6.6    Solving Rational Equations

| Topics |
| --- |
| Solving a New Type of Equation<br>A Visual Interpretation |

**Key Terms**
Use the vocabulary terms listed below to complete each statement in Exercises 1–4.

**clearing fractions        LCD        rational equation        solutions**

1.  A(n) _____ contains one or more rational expressions.

2.  To solve a rational equation is to find all of its _____ .

3.  As a first step in solving rational equations, multiply the equation by the
    _____.

4.  When we multiply an equation by the least common denominator, we are
    _____.

*Solve. If no solution exists, state this.*

5.  $\dfrac{x}{5} + \dfrac{x}{4} = 9$                                    5._____

6.  $\dfrac{2}{5} - \dfrac{1}{t} = \dfrac{1}{2}$                                    6._____

7.  $\dfrac{4}{x+2} + \dfrac{5}{8} = \dfrac{3}{2x+4}$                                    7._____

**8.** $\dfrac{3}{9} - \dfrac{1}{4t} = \dfrac{4}{12}$                  **8.** _____

**9.** $x + \dfrac{10}{x} = 7$                      **9.** _____

**10.** $\dfrac{n+3}{n+1} = \dfrac{2}{n+1}$             **10.** _____

**11.** $\dfrac{t}{t+3} = \dfrac{9}{t^2 + 3t}$            **11.** _____

**12.** $\dfrac{2}{3x+4} = \dfrac{5}{2x}$              **12.** _____

**13.** $\dfrac{2}{x-3} = \dfrac{x}{x+7}$

**13.** _____

**14.** $\dfrac{20}{t-3} - \dfrac{15}{t} = \dfrac{30}{t}$

**14.** _____

**15.** $\dfrac{5}{x-1} - \dfrac{6}{x+2} = \dfrac{3x}{x^2+x-2}$

**15.** _____

**16.** $\dfrac{n}{n+1} - \dfrac{2}{n} = \dfrac{1}{n^2+n}$

**16.** _____

**17.** $\dfrac{4}{x+5} - \dfrac{3}{x-5} = \dfrac{4x}{25-x^2}$

**17.** _____

**18.**  $\dfrac{4}{t^2-8t+16}+\dfrac{t+1}{3t-12}=\dfrac{t}{2t-8}$         **18.** _____

**19.**  $\dfrac{z-1}{z-7}=\dfrac{6}{z-7}$         **19.** _____

**20.**  $\dfrac{6}{y-5}-\dfrac{60}{y^2-25}=1$         **20.** _____

# Chapter 6  RATIONAL EXPRESSIONS AND EQUATIONS

## 6.7    Applications Using Rational Equations and Proportions

**Topics**
> Problems Involving Work
> Problems Involving Motion
> Problems Involving Proportions

**Key Terms**
Use the variables and numbers listed below to complete each equation in Exercises 1 and 2. Variables may be used more than once.

$$a \qquad d \qquad t \qquad 1$$

1.  If $a$ = the time needed for A to complete the work alone, $b$ = the time needed for B to complete the work alone, and $t$ = the time needed for A and B to complete the work together, then

$$\frac{t}{\square} + \frac{\square}{b} = \square .$$

2.  Problems involving motion usually make use of the formula

$$\square = r \cdot \square .$$

*Solve.*

3.  The reciprocal of 2, plus the reciprocal of 5,        3._____
    is the reciprocal of what number?

4.  The sum of a number and 10 times its              4._____
    reciprocal is 7. Find the number.

**5.** The reciprocal of the product of two
consecutive integers is $\dfrac{1}{110}$. Find the two
integers.

5._____

**6.** Mariah can clean the horse stalls at Dazzling
Rides Farm in 6 hr. Lindsay needs 10 hr to
complete the same job. Working together,
how long will it take them to do the job?

6._____

**7.** Circle City's swimming pool can be filled in
15 hr if water enters through a pipe alone or
in 24 hr if water enters through a hose alone.
If water is entering through both the pipe and
the hose, how long will it take to fill the pool?

7._____

**8.** The city records department has a laser jet printer
and an ink jet printer. The ink jet takes three
times the time required by the laser printer to
print an end-of-year summary. If working
together the printers can complete the job in 15
min, how long would it take each printer,
working alone, to print the summary?

8._____

**9.** Alex can detail a car in 20 fewer minutes than it takes Ryan to do the same job. Together, they can detail a car in $\frac{21}{2}$, or $10\frac{1}{2}$ min. How long would it take each person, working alone, to detail a car?

**9.** _____

**10.** Chris can clean an office in 3 hr. When he works with Hannah, they can clean the office in 80 min. How long would it take Hannah, working by herself, to clean the office?

**10.** _____

**11.** Together, Kris and her daughter Kathy require 6 hr 24 min to paint a room. Alone, Kathy would require 24 hr more than Kris. How long would it take Kris to do the job alone? (*Hint*: Convert minutes to hours.)

**11.** _____

**12.** The speed of the current in Pebble Creek is 4 mph. Beth can canoe 6 mi upstream in the same time it takes her to canoe 14 mi downstream. What is the speed of Beth's canoe in still water?

**12.** _____

**13.** The speed of Tom's scooter is 16 mph less than the speed of Mary Lynne's motorcycle. The motorcycle can travel 290 mi in the same time that the scooter can travel 210 mi. Find the speed of each vehicle.

**13.** _____

**14.** Trey's boat travels 24 km/h in still water. He motors 150 km downstream in the same time it takes to go 90 km upstream. What is the speed of the river?

**14.** _____

**15.** A pontoon boat moves 6 km/h in still water. It travels 30 km upriver and 30 km downriver in a total time of 18 hr. What is the speed of the current?

**15.** _____

**16.** A car traveled 220 mi at a certain speed. Had the car been able to travel 15 mph faster, the trip would have been 1.5 hr shorter. How fast did the car go?

**16.** _____

*For each pair of similar triangles, find the value of the indicated letter.*

**17.** $x$                                                 **17.**_____

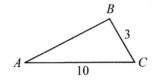 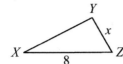

**18.** $p$                                                 **18.**_____

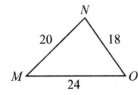 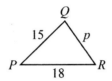

*Use the blueprint below to find the indicated length.*

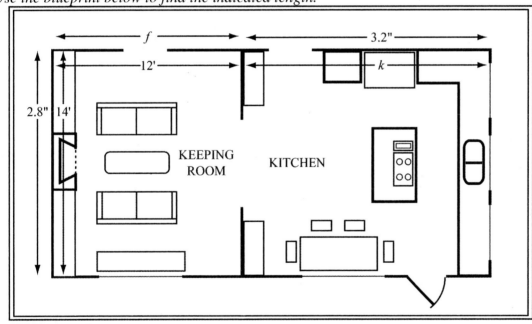

**19.** $f$, in inches on blueprint                       **19.**_____

**20.** $k$, in feet on actual building                   **20.**_____

*Solve.*

**21.** Teri wrote 72 pages for her novel over a period of 12 days. At this rate, how many pages would she write in 16 days?

**21.**_____

**22.** A sample of 48 memory cards contained 3 defective cards. How many defective cards would you expect in a sample of 192 cards?

**22.**_____

**23.** To determine the number of trout in his pond, Oak catches 25 trout, tags them, and lets them loose. Later, he catches 18 trout; 10 of them have tags. Estimate the number of trout in the pond.

**23.**_____

**24.** The ratio of the weight of an object on Saturn to the weight of that object on Earth is 6 to 5.
   **a)** How much would a 12-ton rocket weigh on Saturn?
   **b)** How much would a 150-lb astronaut weigh on Saturn?

**24.a)**_____

**b)**_____

# Chapter 7 SYSTEMS AND MORE GRAPHING

## 7.1    Systems of Equations and Graphing

**Topics**
>   Solutions of Systems
>   Solving Systems of Equations by Graphing

**Key Terms**
Use the vocabulary terms listed below to complete each statement in Exercises 1–6.

|               |              |             |
| ------------- | ------------ | ----------- |
| **consistent**   | **dependent**    | **inconsistent** |
| **independent**  | **intersection** | **system**       |

1.  A(n) _____ of equations is a set of two or more equations, in two or more variables, for which a common solution is sought.

2.  When solving a system of equations graphically, we look for the _____ of the graphs of the equations.

3.  A system of equations that has at least one solution is said to be _____.

4.  A system of equations with no solution is called _____.

5.  If one equation in a system of two equations is a multiple of the other, the equations are _____.

6.  If neither equation in a system of two equations is a multiple of the other, the equations are _____.

*Determine whether the ordered pair is a solution of the given system of equations.*
*Remember to use alphabetical order of variables.*

7.  (2, 1);      $3y - 5 = x$,                     7._____
               $x - 4y = -2$

8.  (3, −1);    $a - 2b = 5$,                      8._____
               $3a - 10 = b$

Name:
Instructor:

Date:
Section:

*Solve each system graphically and check your solution. If a system has an infinite number of solutions, use set-builder notation to write the solution set. If a system has no solution, state this. Where appropriate, round to the nearest hundredth.*

**9.**  $x - y = 1,$

$2y = x + 2$

**9.**_____

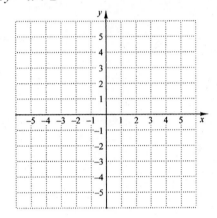

**10.**  $2s - t = 1,$

$t = 3s$

**10.**_____

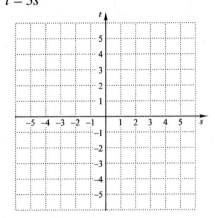

**11.**  $2x = 3y - 2,$

$6y = 5 + 4x$

**11.**_____

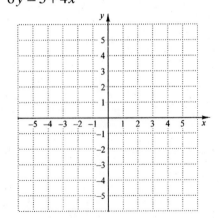

**12.** $y - 3x = 6$,

     $2x + 4 = y$

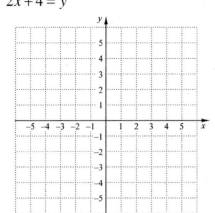

**13.** $y = 2x + 1$,

     $2y - 4x = 2$

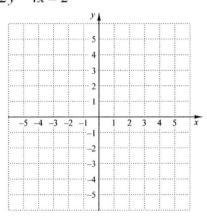

**14.** $y = -2.5x + 8.3$,

     $y = 6.4x - 2.6$

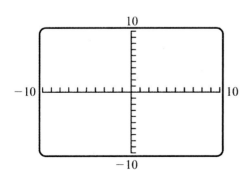

**15.** For the systems in Exercises 9–14, which are consistent?

**15.** _____

**16.** For the systems in Exercises 9 – 14, which contain dependent equations?

**16.** _____

# Chapter 7 SYSTEMS AND MORE GRAPHING

## 7.2    Systems of Equations and Substitution

**Topics**
> The Substitution Method
> Solving for the Variable First
> Problem Solving

*Solve each system using the substitution method. If a system has no solution, state this.*

**1.**  $x + y = 5$,
$y = x + 1$

1._____

**2.**  $x = y + 6$,
$x + 2y = 12$

2._____

**3.**  $y = 3x - 2$,
$2y - x = 1$

3._____

**4.**  $2x + 5y = 9$,
$x = y - 1$

4._____

**5.**  $y = 3x - 1$,
$6x - 2y = 2$

5._____

**6.**  $a + b = -6$,
$\quad b - a = 8$

**6.** _____

**7.**  $4x + 3y = 6$,
$\quad 4x - y = 2$

**7.** _____

**8.**  $6a - 2b = 6$,
$\quad b = 3a - 3$

**8.** _____

**9.**  $3x - y = 3$,
$\quad -5x + y = 3$

**9.** _____

**10.**  $s - 6t = 5$,
$\quad 3s + 2t = 0$

**10.** _____

**11.** $3y + 4 = x$,
$\quad 3y - x = 5$

**11.** _____

**12.** $p - q = -5$,
$\quad 2p = q + 1$

**12.** _____

**13.** $x - 3y = 7$,
$\quad 3x - 5y = 1$

**13.** _____

**14.** $7x + 2y = -4$,
$\quad 8x - y = 2$

**14.** _____

*Solve using a system of equations.*

**15.** The sum of two numbers is 77. One number is 7 more than the other. Find the numbers.

**15.** _____

**16.** Find two numbers for which the sum is 90 and the difference is 14.

**16.**_____

**17.** The difference between two numbers is 16. Five times the larger number is nine times the smaller. What are the numbers?

**17.**_____

**18.** Two angles are supplementary. One angle is 15° less than twice the other. Find the measures of the angles.

**18.**_____

**19.** Two angles are complementary. Their difference is 4°. Find the measure of each angle.

**19.**_____

**20.** A rectangle has a perimeter of 124 ft. The width is 10 ft less than the length. Find the length and the width.

**20.**_____

**21.** The perimeter of a maximum size rugby field is 428 m. The length is 4 m longer than twice the width. Find the dimensions.

**21.**_____

# Chapter 7 SYSTEMS AND MORE GRAPHING

**7.3    Systems of Equations and Elimination**

---

**Topics**

Solving by the Elimination Method

Problem Solving

---

*Solve using the elimination method. If a system has no solution, state this.*

**1.** $x - y = 3$,                                    **1.**_____

$x + y = 11$


**2.** $a + b = 8$,                                    **2.**_____

$-a + 5b = 7$


**3.** $3x - y = 3$,                                   **3.**_____

$-5x + y = 3$


**4.** $6x - 5y = 18$,                                 **4.**_____

$3x + 5y = -4$


**5.** $2x - y = 4$,                                   **5.**_____

$3x + y = -2$

**6.**  $r + s = 9$,
       $3r - 2s = -5$

**6.**_____

**7.**  $8a + 12b = 10$,
       $6a + 9b = 5$

**7.**_____

**8.**  $5r + 8s = 5$,
       $2r + 2s = 1$

**8.**_____

**9.**  $3x - 2y = 5$,
       $6x - 10 = 4y$

**9.**_____

**10.**  $6x - 0.5y = 3$,
        $1.5x + 2.25y = -4$

**10.**_____

**11.**  $0.12s - 0.06t = 12$,
         $0.08s + 0.16t = -16$

**11.**_____

**12.**  $3x + 2y = 2$,
         $9x - 4y = -2$

**12.**_____

*Solve.*

**13.** For a one-day move, Budget rents a 24-ft truck for $39.99 plus 99¢ per mile. Penske rents a 26-ft van for $49.95 plus 79¢ per mile. (*Sources*: Budget Truck Rental; Penske Truck Leasing) For what mileage is the cost the same?

**13.**_____

**14.** Two angles are complementary. One angle is 10° less than four times the other. Find the measure of each angle.

**14.**_____

**15.** Recently, AT&T offered a One-Rate 7¢ Plus Plan costing $3.95 per month plus 7¢ a minute. Their One-Rate Simple Plan has no monthly fee, but costs 29¢ a minute. For what number of minutes will the two plans cost the same?

**15.** _____

**16.** Two angles are supplementary. One angle measures 20° more than four times the measure of the other. Find the measure of each angle.

**16.** _____

**17.** Brittany and Jon packed an order of 840 sports awards. Brittany packed 200 more than Jon. How many did each pack?

**17.** _____

**18.** Easton has 123 ft of fencing for a rectangular garden. If the garden's length is to be $1\frac{1}{2}$ times its width, what should the garden's dimensions be?

**18.** _____

# Chapter 7 SYSTEMS AND MORE GRAPHING

## 7.4    More Applications Using Systems

**Topics**
> Total-Value Problems
> Mixture Problems

**Key Terms**
Use the vocabulary terms listed below to complete each statement in Exercises 1 and 2. The terms may be used more than once.

distance          rate          time
interest          principal     rate

1.  The motion formula $d = rt$ states that _____ equals
    _____ times _____.

2.  The interest formula $I = Prt$ states that _____ equals
    _____ times _____ times _____.

*Solve.*

3.  Together, Sara and Jean ran a total of 96
    miles during the first week of April. Sara ran
    10 more mi than Jean. How many miles did
    each run?

    3._____

4.  Each course at Northside Community College is
    worth either 3 or 4 credits. Five students sharing
    a house are taking a total of 26 courses that are
    worth a total of 92 credits. How many 3-credit
    courses and how many 4-credit courses are being
    taken?

    4._____

**5.** The Clarkstown Volunteer Fire Department served 325 chicken and noodle dinners. A child's plate cost $4.50 and an adult's plate cost $6.00. A total of $1672.50 was collected. How many of each type of plate was served?

5._____

**6.** Admission to Mammoth Cave is $12 for adults and $8 for youth (*Source*: National Park Service). One day, 575 people entered the cave paying a total of $5600. How many adults and how many youth entered the cave?

6._____

**7.** Creative Candles sells 16-oz soy candles for 9.95 and 24-oz candles for $12.95. Last March, they sold 82 candles for a total of $959.90. How many of each size candle were sold?

7._____

**8.** Wholesome Foods sells dried apricots for $12.25 per pound and dried apples for $7.50 per pound. How much of each fruit should be used to make a 40-lb mixture that sells for $9.40 per pound?

8._____

**9.** Alpine Trail Mix is 40% nuts and Meadows Trail Mix is 25% nuts. How much of Alpine and how much of Meadows should be mixed to form a 10-lb batch of trail mix that is 32% nuts?

**9.**_____

**10.** Whitney's two student loans totaled $14,000. One of her loans was at 5% simple interest and the other at 7%. After one year, Whitney owed $850 in interest. What was the amount of each loan?

**10.**_____

**11.** Tropical Punch is 18% fruit juice and Caribbean Spring is 24% fruit juice. How many liters of each should be mixed to get 18 L of a mixture that is 20% juice?

**11.**_____

**12.** A television network plays 40 commercials during a sports game. Each commercial is either 30 sec or 45 sec long. If the total commercial time during the game is 24 min, how many commercials of each type does the network play?

**12.**_____

**13.** Personalized Cell sells graphics for $2.00 each and ringtones for $2.50 each. In one day, a total of $492 was taken in from the sale of a combination of 226 graphics and ringtones. How many of each were sold?

**13.**_____

**14.** Tia makes a $12.45 purchase at the card store with a $20 bill. The store gives her the change in quarters and dimes. There are 38 coins in all. How many of each kind are there?

**14.**_____

# Chapter 7 SYSTEMS AND MORE GRAPHING

### 7.5    Linear Inequalities in Two Variables

---

**Topics**
>   Graphing Linear Inequalities
>   Linear Inequalities in One Variable

---

**Key Terms**
Use the vocabulary terms listed below to complete each statement in Exercises 1–2.

> **linear inequality**                          **test point**

1.   The sentence $2x - 3 \leq -5$ is an example of a(n) _____.

2.   To determine which side of the boundary to shade as the graph of the solution set of an inequality, select a(n) _____ not on the line.

*Determine whether each ordered pair is a solution of the given inequality.*
3.   $(2, -5)$ ; $2x + 4y \leq -5$                          3._____

4.   $(20, 12)$ ; $2y - 3x > 1$                          4._____

*Graph on a plane.*

**5.** $y \leq \frac{1}{3}x$

**5.**

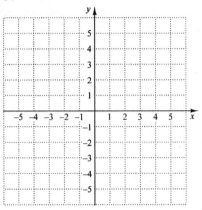

**6.** $y \geq x - 2$

**6.**

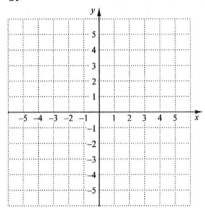

**7.** $4x - y > 8$

**7.**

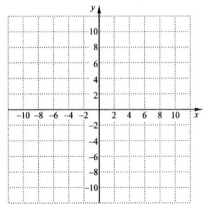

**8.**     $x - 3y < 6 - x$

**8.**

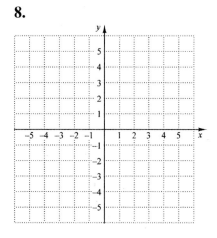

**9.**     $-3 \le y \le 4$

**9.**

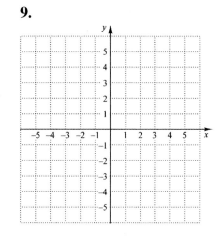

**10.**    $0 \le x \le 5$

**10.**

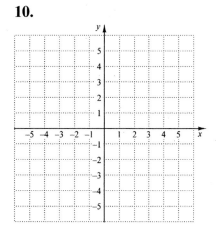

**11.**    $y < x - 4$

**11.**

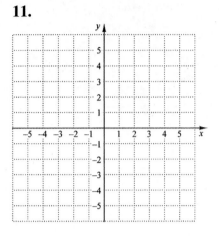

**12.**    $5x - 3y \geq 18$

**12.**

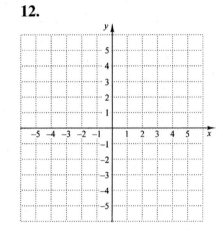

**13.**    $3y > -12$

**13.**

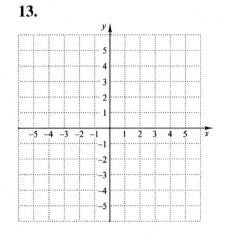

# Chapter 7 SYSTEMS AND MORE GRAPHING

## 7.6    Systems of Linear Inequalities

**Topics**
> Graphing Systems of Inequalities
> Locating Solution Sets

*Graph the solutions of each system.*

**1.**    $y < -x$,
    $y > x - 3$

**1.**

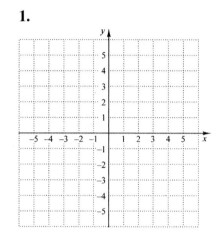

**2.**    $y \le 2x$,
    $y \ge 4x - 3$

**2.**

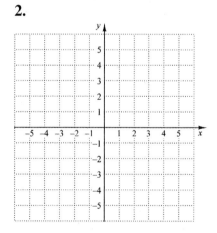

**3.**  $y \geq -3$,
     $y \leq 2 - x$

**3.**

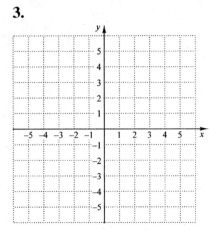

**4.**  $2x + y \geq 1$,
     $2x + y \leq 4$

**4.**

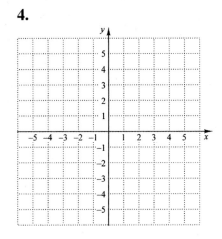

**5.**  $y \geq x - 6$,
     $y \leq 3x - 4$,
     $x \leq 5$

**5.**

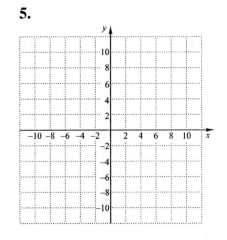

Name:                          Date:
Instructor:                    Section:

**6.**    $2x + 4y \le 24$,
          $5x + 3y \le 30$,
          $x \ge 0$,
          $y \ge 0$

**6.**

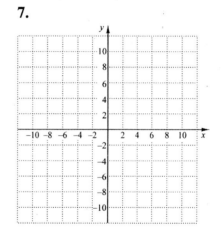

**7.**    $-4 > x - 2y$
          $x + 6y < 36$
          $2y \ge -3x - 12$

**7.**

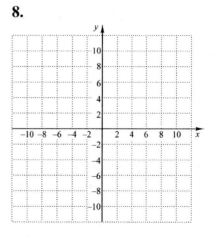

**8.**    $y - 4 > 2x$
          $y > -6x$
          $3y - 18 > x$

**8.**

**9.**   $y \leq -\dfrac{2}{3}x + 5$

$y < 2x + 4$

$x < 4$

$y \geq -1$

**9.**

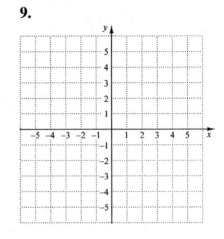

**10.**   $y < 6x + 18$

$y \geq 6x - 18$

$y < -\dfrac{1}{2}x + 2$

$y \geq -\dfrac{1}{2}x - 4$

**10.**

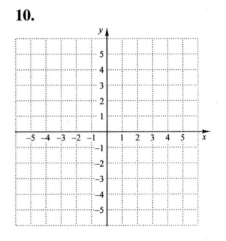

# Chapter 7 SYSTEMS AND MORE GRAPHING

## 7.7    Direct and Inverse Variation

| Topics |
|---|
| Equations of Direct Variation |
| Problem Solving with Direct Variation |
| Equations of Inverse Variation |
| Problem Solving with Inverse Variation |

## Key Terms

Use the vocabulary terms listed below exactly once to complete each statement in Exercises 1–2.

**direct**                    **inverse**

1. _____ variation is described by a function of the form $f(x) = kx$.

2. _____ variation is described by a function of the form $f(x) = \dfrac{k}{x}$.

*Find an equation of variation in which y varies directly as x and the following are true.*

3.    $y = 55$ when $x = 5$                    3._____

4.    $y = \frac{2}{3}$ when $x = 8$                    4._____

*Find an equation of variation in which y varies inversely as x and the following are true.*

5.    $y = 7$ when $x = 9$                    5._____

Name:

Instructor:

Date:

Section:

**6.**    $y = 72$ when $x = \frac{1}{12}$

**6.**_____

*Solve.*

**7.**    Hooke's law states that the distance $d$ that a spring is stretched by a hanging object varies directly as the mass $m$ of the object. If the distance is 50 cm when the mass is 6 kg, what is the distance when the mass is 2 kg?

**7.**_____

**8.**    The electric current $I$, in amperes, in a circuit varies directly as the voltage $V$. When 12 volts are applied, the current is 3 amperes. What is the current when 16 volts are applied?

**8.**_____

**9.**    The time $T$ required to do a job varies inversely as the number of people $P$ working. It takes 4 hr for 9 people to weed the community garden. How long would it take 10 people to complete the job?

**9.**_____

**10.** The number of calories burned by a person in a Zumba aerobic class is directly proportional to the time spent exercising. It takes 10 min to burn 80 calories (*Source*: Family Fun and Fitness). How long would it take to burn 200 calories in the class?

**10.**_____

**11.** The wavelength $W$ of a radio wave varies inversely as its frequency $F$. A wave with a frequency of 1600 kilohertz has a length of 225 meters. What is the length of a wave with a frequency of 3000 kilohertz?

**11.**_____

**12.** The current $I$ in an electrical conductor varies inversely as the resistance $R$ in the conductor. If the current is $\frac{2}{5}$ ampere when the resistance is 120 ohms, what is the current when the resistance is 150 ohms?

**12.**_____

*Find each equation of variation.*

**13.**   $y$ varies directly as the square of $x$, and           **13.**_____
       $y = 90$ when $x = 6$.

**14.**   $y$ varies inversely as the square of $x$, and      **14.**_____
       $y = 72$ when $x = 0.5$.

*Solve.*

**15.**   The intensity $I$ of a wireless signal varies      **15.**_____
       inversely as the square of the distance $d$ from
       the transmitter. If the intensity is 50 W/m$^2$ at
       a distance of 5 km, what is the intensity
       20 km from the transmitter?

# Chapter 8 RADICAL EXPRESSIONS AND EQUATIONS

## 8.1    Introduction to Square Roots and Radical Expressions

**Topics**
> Square Roots
> Radicands and Radical Expressions
> Irrational Numbers
> Square Roots and Absolute Value
> Problem Solving

**Key Terms**
Use the vocabulary terms listed below to complete each statement in Exercises 1–3.

|  |  |  |
|---|---|---|
| **irrational** | **principal** | **radicand** |

1.  The _____ square root of a nonnegative number is its nonnegative square root.

2.  The square root of any whole number that is not a perfect square is a(n) _____ number.

3.  The _____ is the expression under the radical sign..

*Classify each statement as either true or false. Do not use a calculator*

4.  $\sqrt{111}$ is between 10 and 11.                4._____

5.  $\sqrt{57}$ is between 6 and 7.                5._____

*For each number, find all of its square roots.*
6.  16                                       6._____

7.  64                                       7._____

8.  4900                                     8._____

*Simplify.*

**9.** $-\sqrt{64}$

9._____

**10.** $\sqrt{256}$

10._____

**11.** $\sqrt{0.09}$

11._____

**12.** $-\sqrt{0.0121}$

12._____

**13.** $\sqrt{400}$

13._____

**14.** $-\sqrt{484}$

14._____

*Identify the radicand for each expression.*

**15.** $7\sqrt{x+1}+5$

15._____

**16.** $\sqrt{\dfrac{2s}{3t}}$

16._____

*Classify each number as rational or irrational;*

**17.** $\sqrt{49}$

17._____

**18.** $\sqrt{123}$

18._____

*Use a calculator to approximate each of the following numbers. Round to three decimal places.*

**19.** $\sqrt{2}$

                                         **19.**_____

**20.** $\sqrt{61}$

                                         **20.**_____

*Simplify. Assume that x can represent any real number.*

**21.** $\sqrt{4x^2}$

                                         **21.**_____

**22.** $\sqrt{(8-x)^2}$

                                         **22.**_____

**23.** $\sqrt{(x+7)^2}$

                                         **23.**_____

*Simplify. Assume that all variables represent nonnegative numbers.*

**24.** $\sqrt{9x^2}$

                                         **24.**_____

**25.** $\sqrt{(31d)^2}$

                                         **25.**_____

**26.** $\sqrt{(x+5)^2}$

                                         **26.**_____

**27.** A parking lot has attendants to park cars, and it uses spaces where cars are left before they are taken to permanent parking stalls. The number $N$ of such spaces needed is approximated by the formula $N = 2.5\sqrt{A}$, where $A$ is the average number of arrivals in peak hours. Find the number of spaces needed when the average number of arrivals is

**a.** 16

**b.** 31

**27a.**_____

**b.**_____

**28.** An athlete's hang time (time airborne for a jump), $T$, in seconds, is given by $T = 0.144\sqrt{V}$, where $V$ is the athlete's vertical leap, in inches. A basketball player can jump 16 inches vertically. Find the player's hang time.

**28.**_____

# Chapter 8 RADICAL EXPRESSIONS AND EQUATIONS

### 8.2     Multiplying and Simplifying Radical Expressions

| **Topics** |
| --- |
| Multiplying |
| Simplifying and Factoring |
| Simplifying Square Roots of Powers |
| Multiplying and Simplifying |

*Multiply. Assume all variables represent nonnegative numbers.*

**1.** $\sqrt{10}\sqrt{11}$

1._____

**2.** $\sqrt{13}\sqrt{7}$

2._____

**3.** $\sqrt{\dfrac{11}{5}}\sqrt{\dfrac{3}{22}}$

3._____

**4.** $\sqrt{20}\sqrt{20}$

4._____

**5.** $\sqrt{49}\sqrt{5}$

5._____

**6.** $\sqrt{5}\sqrt{2p}$

6._____

**7.** $\sqrt{10p}\sqrt{5q}$

7._____

*Simplify by factoring. Assume that all variables represent nonnegative numbers.*

**8.** $\sqrt{175}$

8._____

**9.** $\sqrt{90}$

9._____

**10.** $\sqrt{720}$

**11.** $\sqrt{36x}$

**12.** $\sqrt{50x}$

**13.** $\sqrt{25x^2}$

**14.** $\sqrt{27t^2}$

*Evaluate $\sqrt{b^2 - 4ac}$ for the values of a, b, and c given.*
**15.** $a = 7, b = 4, c = -6$

**16.** $a = -8, b = -8, c = 4$

*Simplify. Assume that all variables represent nonnegative numbers.*
**17.** $\sqrt{x^4}$

**18.** $\sqrt{x^{11}}$

_____

**19.** $\sqrt{t^9}$

**19.**_____

**20.** $\sqrt{18a^9}$

**20.**_____

*Multiply and, if possible, simplify.*
**21.** $\sqrt{14}\sqrt{21}$

**21.**_____

**22.** $\sqrt{6}\sqrt{10}$

**22.**_____

**23.** $\sqrt{5x}\sqrt{45y}$

**23.**_____

**24.** $\sqrt{5b}\sqrt{15b}$

**24.**_____

**25.** $\sqrt{ab}\sqrt{ac}$

25._____

**26.** $\sqrt{11q}\sqrt{121q^3}$

26._____

**27.** $\sqrt{28x^5}\sqrt{28x^5}$

27._____

**28.** $\sqrt{a^4b^7}\sqrt{ab^2}$

28._____

*The formula $r = 2\sqrt{5L}$ can be used to approximate the speed, r, in miles per hour, of a car that has left a skid mark L feet long.*

**29.** What was the speed of a car that left skid marks of 200 ft? of 30 ft? Round answers to the nearest tenth.

29._____

**30.** What was the speed of a car that left skid marks of 180 ft? of 230 ft? Round answers to the nearest tenth.

30._____

# Chapter 8 RADICAL EXPRESSIONS AND EQUATIONS

## 8.3    Quotients Involving Square Roots

**Topics**
> Dividing Radical Expressions
> Rationalizing Denominators

**Key Terms**
Use the vocabulary terms listed below to complete each statement in Exercises 1 and 2.

**multiply by 1**                 **rationalizing**

1. When we find an equivalent expression in which the denominator no longer contains a radical, we are _____ the denominator.

2. To make the denominator of the radicand a perfect power, we _____ under the radical.

*Simplify. Assume that all variables represent positive numbers.*

3.    $\sqrt{\dfrac{81}{49}}$                               3._____

4.    $\dfrac{\sqrt{45}}{\sqrt{5}}$                               4._____

5.    $\dfrac{\sqrt{30}}{\sqrt{15}}$                               5._____

6.    $\dfrac{\sqrt{2}}{\sqrt{18}}$                               6._____

**7.** $\dfrac{\sqrt{75y^3}}{\sqrt{3y}}$

**7.** _____

**8.** $\dfrac{\sqrt{30x}}{\sqrt{5x}}$

**8.** _____

**9.** $\dfrac{\sqrt{18d^{13}}}{\sqrt{2d}}$

**9.** _____

**10.** $\sqrt{\dfrac{16p^5}{q^{10}}}$

**10.** _____

*Form an equivalent expression by rationalizing each denominator. Assume that all variables represent positive numbers.*

**11.** $\dfrac{5}{\sqrt{3}}$

**11.** _____

**12.** $\dfrac{\sqrt{2}}{\sqrt{5}}$

**12.** _____

**13.** $\dfrac{\sqrt{4}}{\sqrt{125}}$

**13.** _____

**14.** $\dfrac{\sqrt{m}}{\sqrt{27}}$

**14.** _____

**15.** $\dfrac{3\sqrt{7}}{5\sqrt{2}}$

**15.** _____

**16.** $\sqrt{\dfrac{6}{x}}$

**16.** _____

**17.** $\dfrac{\sqrt{y}}{\sqrt{12}}$

**17.** _____

18.    $\sqrt{\dfrac{8}{50x^2 y}}$                                        18._____

19.    $\dfrac{\sqrt{6a^5 b}}{\sqrt{2a}}$                                       19._____

20.    $\dfrac{\sqrt{81xyz}}{3\sqrt{3}}$                                       20._____

*The period T of a pendulum is the time it takes to move from one side to the other and back. A formula for the period is $T = 2\pi\sqrt{\dfrac{L}{32}}$, where T is in seconds and L is in feet. Use 3.14 for $\pi$.*

21.    Find the periods of pendulums of lengths 50 ft and 8 ft.                        21._____

22.    The pendulum of a grandfather clock is $\dfrac{32}{\pi^2}$ ft long. How long does it take to complete one swing back and forth?         22._____

# Chapter 8 RADICAL EXPRESSIONS AND EQUATIONS

## 8.4    Radical Expressions with Several Terms

| **Topics** |
| --- |
| Adding and Subtracting Radical Expressions |
| More with Multiplication |
| More with Rationalizing Denominators |

**Key Terms**

Use the vocabulary terms listed below to complete each statement in Exercises 1 and 2.

**conjugates**                          **like radicals**

1.    When two radical expressions have the same indices and radicands, they are called

_____.

2.    Pairs of radical terms like $\sqrt{a} + \sqrt{b}$ and $\sqrt{a} - \sqrt{b}$ are called _____.

*Add or subtract. Simplify by combining like radical terms, if possible. Assume that all variables and radicands represent positive real numbers.*

3.   $5\sqrt{11} - 8\sqrt{11}$                    3._____

4.   $2\sqrt{2} + 8\sqrt{2}$                      4._____

5.   $4\sqrt{7} - 2\sqrt{7}$                      5._____

6.   $2\sqrt{d} - 5\sqrt{d}$                      6._____

7.   $6\sqrt{18} - 5\sqrt{8}$                     7._____

8.   $4\sqrt{12} + 9\sqrt{3}$                     8._____

**9.** $\sqrt{63} - 2\sqrt{7}$

9._____

**10.** $5\sqrt{12} + \sqrt{48} - \sqrt{8}$

10._____

**11.** $6\sqrt{8} + \sqrt{32} - \sqrt{50}$

11._____

**12.** $\sqrt{36z} + \sqrt{49z} - 11\sqrt{z}$

12._____

*Multiply. Assume all variables represent nonnegative real numbers.*

**13.** $2\sqrt{3}\left(\sqrt{3} - \sqrt{10}\right)$

13._____

**14.** $\sqrt{5}(\sqrt{11} - \sqrt{3})$

14._____

**15.** $\left(3 + \sqrt{5}\right)\left(7 - \sqrt{5}\right)$

15._____

**16.** $\left(2\sqrt{5} + 3\sqrt{2}\right)\left(3\sqrt{5} - 6\sqrt{2}\right)$

16._____

**17.** $\left(\sqrt{2x} - \sqrt{7}\right)^2$

17._____

**18.**  $\left(1+\sqrt{t-7}\right)^2$  **18.**_____

*Rationalize each denominator. If possible, simplify to form an equivalent expression.*

**19.**  $\dfrac{2}{5+\sqrt{3}}$  **19.**_____

**20.**  $\dfrac{9}{1+\sqrt{2}}$  **20.**_____

**21.**  $\dfrac{\sqrt{p}}{\sqrt{p}-\sqrt{q}}$  **21.**_____

**22.**  $\dfrac{2}{7-\sqrt{6}}$  **22.**_____

**23.**  $\dfrac{6}{\sqrt{14}+1}$  **23.**_____

**24.**  $\dfrac{\sqrt{5}}{\sqrt{5}-6}$  **24.**_____

**25.**  $\dfrac{\sqrt{7}}{\sqrt{7}-\sqrt{2}}$

25._____

**26.**  $\dfrac{\sqrt{22}}{\sqrt{10}+\sqrt{22}}$

26._____

**27.**  $\dfrac{3}{\sqrt{19}-\sqrt{2}}$

27._____

**28.**  $\dfrac{\sqrt{7}-\sqrt{6}}{\sqrt{7}+\sqrt{6}}$

28._____

**29.**  $\dfrac{6+\sqrt{7}}{2-\sqrt{5}}$

29._____

**30.**  $\dfrac{2\sqrt{11}-4\sqrt{5}}{6\sqrt{2}+3\sqrt{5}}$

30._____

# Chapter 8 RADICAL EXPRESSIONS AND EQUATIONS

## 8.5     Radical Equations

**Topics**
> Solving Radical Equations
> Problem Solving and Applications

**Key Terms**
Use the vocabulary terms listed below to complete each statement in Exercises 1 and 2.

**principle of squaring**                    **radical equation**

1.  A_____ is an equation in which the variable appears in a radicand.

2.  The _____ states that if $a = b$, then $a^2 = b^2$.

*Solve.*

3.  $\sqrt{x} = 8$                                              3._____

4.  $\sqrt{3x-7} = 5$                                      4._____

5.  $\sqrt{6x} - 3 = 7$                                     5._____

6.  $\sqrt{n-8} - 2 = 1$                                  6._____

7. $\sqrt{z+2}+6=10$

7._____

8. $\sqrt{2x+8}=11$

8._____

9. $7\sqrt{a}=a$

9._____

10. $5+\sqrt{x-9}=7$

10._____

11. $30-6\sqrt{5n}=0$

11._____

12. $\sqrt{5x-11}=\sqrt{x+10}$

12._____

13. $\sqrt{2y+4}=\sqrt{2y-5}$

13._____

**14.** $x - 4 = \sqrt{x - 2}$

14._____

**15.** $2\sqrt{x - 2} = x - 2$

15._____

**16.** $\sqrt{7x + 46} = x + 4$

16._____

**17.** $\sqrt{6z + 1} = \sqrt{5z + 3}$

17._____

**18.** $x = 1 + 3\sqrt{x - 3}$

18._____

**19.** $7 + 2\sqrt{x + 1} = x$

19._____

**20.** $\sqrt{5x + 1} = 1 + \sqrt{4x - 3}$

20._____

**21.** $7+\sqrt{10-z}=8+\sqrt{1-z}$

**22.** $2\sqrt{3y-2}-\sqrt{4y-3}=1$

**23.** $\sqrt{(p+8)(p+3)}-4=p+1$

**24.** The formula $r=2\sqrt{5L}$ can be used to approximate the speed $r$, in miles per hour, of a car that has left a skid mark of length $L$, in feet. How far will a car skid at 20 mph? at 30 mph? Round to the nearest tenth.

**25.** A steeplejack can see 14 km to the horizon from the top of a building. These numbers are related by the equation $V=3.5\sqrt{h}$, where $V$ is the distance in kilometers from the steeplejack's eyes to the horizon and $h$ is the height in meters of the steeplejack's eyes. What is the altitude (to the nearest meter) of the steeplejack's eyes?

Name:                                    Date:
Instructor:                              Section:

# Chapter 8 RADICAL EXPRESSIONS AND EQUATIONS

**8.6    Applications Using Right Triangles**

**Topics**
>    Right Triangles
>    Problem Solving
>    The Distance Formula

*In a right triangle, find the length of the side not given. If an answer is not a whole number, use radical notation to give the exact answer and decimal notation for an approximation to the nearest thousandth.*

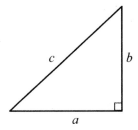

**1.**  $a = 3, b = 3$                       1._____

**2.**  $a = 9, c = 15$                      2._____

**3.**  $a = 4\sqrt{3}, c = 8$                 3._____

**4.**  $a = 6, c = 10$                       4._____

**5.**  $a = 4, b = 7$                        5._____

**6.** $a = 6, b = 6$

**6.**_____

**7.** $a = 8, c = 17$

**7.**_____

**8.** $b = 5, c = 20$

**8.**_____

*In Exercises 9–14, give an exact answer and, where appropriate, an approximation to the nearest thousandth.*

**9.** A right triangle's hypotenuse is 6 m and one leg is $2\sqrt{5}$ m. Find the length of the other leg.

**9.**_____

**10.** The hypotenuse of a right triangle is $\sqrt{40}$ ft and one leg measures 5 ft. Find the length of the other leg.

**10.**_____

**11.** One leg of a right triangle is 5 cm and the hypotenuse measures $5\sqrt{2}$ cm. Find the length of the other leg.

**11.**_____

**12.** A diagonal sidewalk cuts across a rectangular grassy area between classroom buildings. The area is 400 ft long and 300 ft wide. How long is the sidewalk?

**12.**_____

**13.** How long is a guy wire if it reaches from the top of a 25-ft pole to a point on the ground 10 ft from the pole?

**13.**_____

**14.** A laptop has a 17-in. diagonal screen. The height of the screen is 9 in. What is its width?

**14.**_____

**15.** A 13-ft ladder is placed against a vertical wall of a building, with the bottom of the ladder standing on level ground 9 feet from the base of the building. How high up the wall does the ladder reach?

**15.**_____

**16.** A new water pipe is being prepared so that it will run diagonally under a kitchen floor. If the kitchen is 18 ft wide and 27 ft long, how long should the pipe be?

**16.**_____

**17.** The length and the width of a rectangle are given by consecutive integers. The area of the rectangle is 110 m$^2$. Find the length of the diagonal of the rectangle.

**17.**_____

**18.** A soccer field is 80 yd wide and 130 yd long. Find the length of a diagonal of such a field.

**18.**_____

*Find the distance between each pair of points. If an answer is not a whole number, use radical notation to give the exact answer and decimal notation for an approximation to the nearest thousandth.*

**19.**    (2,6) and (4,−3)

**19.**_____

**20.**    (−1,5) and (−2,7)

**20.**_____

**21.**    (3,−8) and (0,−5)

**21.**_____

**22.**    (−4,7) and (6,−1)

**22.**_____

# Chapter 8 RADICAL EXPRESSIONS AND EQUATIONS

## 8.7    Higher Roots and Rational Exponents

**Topics**

Higher Roots
Products and Quotients Involving Higher Roots
Rational Exponents
Calculators

**Key Terms**

In Exercises 1–3, match each expression with an equivalent expression in the right column.

1. _____ $a^{-n}$

    **a)** $\sqrt[n]{a^m}$

2. _____ $a^{m/n}$

    **b)** $\dfrac{1}{a^n}$

3. _____ $a^m \cdot a^n$

    **c)** $a^{m+n}$

Use the vocabulary terms listed below to complete each statement in Exercises 4–6.

    **cube**            **index**            **square**

4. The number $c$ is a(n) _____ root of $a$ if $c^2 = a$.

5. The number $c$ is a(n) _____ root of $a$ if $c^3 = a$.

6. The number $n$ in the expression $\sqrt[n]{a}$ is called the _____.

*Note: Assume for all exercises that even roots are of nonnegative quantities and that all denominators are nonzero.*

*Simplify. If an expression does not represent a real number, state this.*

7. $\sqrt[3]{-\dfrac{8}{27}}$

                              **7.**_____

**8.** $\sqrt[6]{(5x)^6}$

**8.** _____

**9.** $-\sqrt[3]{125}$

**9.** _____

**10.** $\sqrt[4]{10,000}$

**10.** _____

**11.** $\sqrt[3]{-8t^3}$

**11.** _____

**12.** $\sqrt[3]{\dfrac{125}{64}}$

**12.** _____

**13.** $\sqrt[3]{\dfrac{1000x^7}{27y^3}}$

**13.** _____

**14.** $\sqrt[4]{\dfrac{a^8b^7}{c^{14}}}$

**14.** _____

**15.** $\sqrt[6]{\dfrac{m^7 n^6}{p^{20}}}$

**15.** _____

*Simplify.*

**16.** $1000^{1/3}$

**16.** _____

**17.** $8^{5/3}$

**17.** _____

**18.** $4096^{1/6}$

**18.** _____

**19.** $343^{1/3}$

**19.** _____

**20.** $625^{1/4}$

**20.** _____

**21.** $243^{3/5}$

**21.** _____

**22.** $16^{3/4}$

**22.** _____

**23.** $27^{5/3}$                      **23.** _____

**24.** $512^{2/3}$                     **24.** _____

**25.** $27^{-2/3}$                     **25.** _____

**26.** $36^{3/2}$                      **26.** _____

**27.** $512^{-1/3}$                    **27.** _____

**28.** $32^{-1/5}$                     **28.** _____

**29.** $8^{-4/3}$                      **29.** _____

**30.** $729^{-5/3}$                    **30.** _____

# Chapter 9 QUADRATIC EQUATIONS

## 9.1     Solving Quadratic Equations: The Principle of Square Roots

---

**Topics**
> The Principle of Square Roots
> Solving Quadratic Equations of the Type $(x + k)^2 = p$

---

*Solve. Use the principle of square roots.*

1. $x^2 = 196$                              1._____

2. $x^2 = 4$                                   2._____

3. $c^2 = 2$                                   3._____

4. $3x^2 = 12$                               4._____

5. $6t^2 = 18$                               5._____

6. $3x^2 = 39$                               6._____

7. $49 - 81x^2 = 0$                       7._____

**8.** $\quad 4x^2 - 81 = 0$

**8.**_____

**9.** $\quad 64y^2 - 2 = 32$

**9.**_____

**10.** $\quad (y-3)^2 = 100$

**10.**_____

**11.** $\quad 3x^2 - 7 = 0$

**11.**_____

**12.** $\quad (x-5)^2 = 2$

**12.**_____

**13.** $\quad (x-5)^2 = 4$

**13.**_____

**14.** $\quad (x+4)^2 = 25$

**14.**_____

**15.** $\quad (x+5)^2 = 6$

**15.**_____

**16.** $(a+7)^2 = 4$

**16.** _____

**17.** $(x-9)^2 = 49$

**17.** _____

**18.** $(3-x)^2 = 12$

**18.** _____

**19.** $(x+2)^2 = 25$

**19.** _____

**20.** $\left(y+\frac{2}{3}\right)^2 = \frac{8}{9}$

**20.** _____

**21.** $\left(x-\frac{2}{7}\right)^2 = \frac{5}{7}$

**21.** _____

**22.** $x^2 - 4x + 4 = 16$

**22.** _____

**23.**   $x^2 + 14x + 49 = 9$

**23.** _____

**24.**   $v^2 - 14v + 49 = 3$

**24.** _____

**25.**   $x^2 + 4x + 4 = 16$

**25.** _____

# Chapter 9 QUADRATIC EQUATIONS

## 9.2 Solving Quadratic Equations: Completing the Square

---

**Topics**
>    Completing the Square
>    Solving by Completing the Square

---

*Replace the blanks in each equation with constants to complete the square and form a true equation.*

**1.** $x^2 + 20x + \underline{\phantom{xx}} = (x + \underline{\phantom{xx}})^2$

1._____

**2.** $t^2 - 12t + \underline{\phantom{xx}} = (t - \underline{\phantom{xx}})^2$

2._____

**3.** $t^2 - 5t + \underline{\phantom{xx}} = (t - \underline{\phantom{xx}})^2$

3._____

**4.** $x^2 + \frac{4}{7}x + \underline{\phantom{xx}} = (x + \underline{\phantom{xx}})^2$

4._____

*Determine the number that will complete the square. Check by multiplying.*

**5.** $t^2 - 18t$

5._____

**6.** $y^2 + 8y$

6._____

**7.** $t^2 - 7t$

**8.** $m^2 + 11m$

**9.** $n^2 - \dfrac{5}{6}n$

**10.** $y^2 + \dfrac{7}{6}y$

*Solve by completing the square.*
**11.** $t^2 - 4t + 2 = 0$

**12.** $x^2 + 6x - 2 = 0$

**13.** $x^2 + 5x - 6 = 0$

**14.** $x^2 - 14x - 4 = 0$                    **14.**_____

**15.** $3x^2 - 54x - 12 = 0$                  **15.**_____

**16.** $x^2 - 6x - 4 = 0$                     **16.**_____

**17.** $x^2 + 2x - 4 = 0$                     **17.**_____

**18.** $2x^2 + 14x - 60 = 0$                  **18.**_____

**19.** $x^2 + x = 20$                         **19.**_____

**20.** $t^2 - 6t = -8$                        **20.**_____

**21.** $x^2 + \dfrac{13}{2}x - 13 = 0$

**21.** _____

**22.** $2x^2 - x - 17 = 0$

**22.** _____

**23.** $2x^2 + 5x - 1 = 0$

**23.** _____

**24.** $2x^2 = 13x + 5$

**24.** _____

**25.** $8x^2 + 3 = 10x$

**25.** _____

**26.** $3x^2 + 2x - 4 = 0$

**26.** _____

# Chapter 9 QUADRATIC EQUATIONS

## 9.3    The Quadratic Formula and Applications

| Topics |
| --- |
| The Quadratic Formula |
| Problem Solving |

**Key Terms**

In Exercises 1 and 2, complete each statement.

1.    The principle of square roots states that for any real number $k$, if $x^2 = k$, then

   $x =$ _____ .

2.    The quadratic formula states that the solutions of $ax^2 + bx + c = 0, a \neq 0$, are given by

   $x =$ _____ .

*Solve. If no real-number solutions exist, state this.*

3.    $x^2 + 3x - 2 = 0$                          3._____

4.    $4a^2 = 32a - 20$                            4._____

5.    $x^2 + x + 3 = 0$                            5._____

**6.** $t^2 + 11 = 4t$

**6.**_____

**7.** $\dfrac{1}{x^2} - 2 = \dfrac{8}{x}$

**7.**_____

**8.** $4s + s(s-3) = 1$

**8.**_____

**9.** $10x^2 + 5x = 2$

**9.**_____

**10.** $100x^2 - 60x + 9 = 0$

**10.**_____

**11.** $6x(x+1)+3=5x(x+2)$          **11.** _____

**12.** $5t^2=12t+16$          **12.** _____

**13.** $x^2+25=6x$          **13.** _____

**14.** $2y^2-3y=1$          **14.** _____

**15.** $5x^2-9x-17=0$          **15.** _____

**16.** $3t^2-7t=3$          **16.** _____

*Solve using the quadratic formula. Use a calculator to approximate the solutions to the nearest thousandth.*

**17.** $x^2 + 2x - 5 = 0$                    **17.**_____

**18.** $x^2 - 5x + 2 = 0$                    **18.**_____

**19.** $2x^2 + 4x - 7 = 0$                   **19.**_____

**20.** $3x^2 - 5x - 4 = 0$                   **20.**_____

*Solve.*

**21.** A polygon has 20 diagonals. How many sides        **21.**_____
does it have? (Use $d = (n^2 - 3n)/2$ .)

22. The height of a building is 1439 ft. How long would it take an object to fall to the ground from the top? (Use $s = 16t^2$, where $t$ is in seconds; round answer to the nearest tenth.)

22._____

23. The hypotenuse of a right triangle is 39 ft long. The length of one leg is 21 ft more than the other. Find the lengths of the legs.

23._____

24. The length of a rectangle is 2 m greater than the width. The area is 120 m². Find the length and the width.

24._____

25. The diagonal of a rectangular room is 13 ft long. One wall measures 7 ft longer than the adjacent wall. Find the dimensions of the room.

25._____

**26.** The area of a right triangle is 10 m$^2$. One leg is 19 m longer than the other. Find the lengths of the legs.

**26.**_____

*The formula $A = P(1+r)^t$ is used to find the value A to which P dollars grows when invested for t years at an annual interest rate r.*

**27.** Find the interest rate if $2560 grows to $3613.04 in 2 years. Round to the nearest tenth of a percent.

**27.**_____

**28.** Find the interest rate if $2560 grows to $2892.72 in 2 years. Round to the nearest tenth of a percent.

**28.**_____

# Chapter 9 QUADRATIC EQUATIONS

## 9.4    Formulas

| **Topics** |
| --- |
| Solving Formulas |

*Solve each formula for the specified variable.*

**1.** $A = \dfrac{1}{9}vt$, for $v$

1._____

**2.**  $R = 4\pi zd$, for $z$

2._____

**3.**  $P = 14\sqrt{C}$, for $C$

3._____

**4.**  $\dfrac{1}{Y} = \dfrac{1}{t} + \dfrac{1}{m}$, for $Y$

4._____

**5.**  $S = vA^2 + uA$, for $A$
(Use only the positive square root.)

5._____

**6.** $\dfrac{1}{100} = \dfrac{a-5}{M}$, for $a$

**6.** _____

**7.** $T = 8v^2$, for $v$
(Use only the positive square root.)

**7.** _____

**8.** $y = 3x^2 + 5x$, for $x$
(Use only the positive square root.)

**8.** _____

**9.** $y = \dfrac{kxz}{w^2}$, for $w$
(Use only the positive square root.)

**9.** _____

**10.** $M = 5\sqrt{\dfrac{p}{q}}$, for $q$

**10.** _____

**11.** $a^2 = b^2 + c^2 - 2abX$, for $c$
(Use only the positive square root.)

**11.** _____

**12.**   $t = x_0 v + \dfrac{av^2}{8}$ , for $v$                    **12.** _____

(Use only the positive square root.)

**13.**   $B = \frac{1}{3}(x^2 - 2x)$ , for $x$                    **13.** _____

(Use only the positive square root.)

**14.**   $H = 1.6\sqrt{d}$ , for $d$                    **14.** _____

**15.**   $Cf = Mf - e$, for $f$                    **15.** _____

**16.**   $Bx + Gy = K$, for $y$                    **16.** _____

**17.**   $\dfrac{p}{w} = \dfrac{r}{p}$, for $p$                    **17.** _____

**18.**   $vm^2 + wm + x = 0,$ for $m$

**18.** _____

**19.**   $\dfrac{D}{K} = N,$ for $K$

**19.** _____

**20.**   $m - g = \dfrac{h}{m},$ for $m$

**20.** _____

**21.**   $d = \sqrt{a^2 - f^2},$ for $a$
       (Use only the positive square root.)

**21.** _____

**22.**   $\sqrt{6m - g} = \sqrt{h + m},$ for $m$

**22.** _____

# Chapter 9 QUADRATIC EQUATIONS

### 9.5    Complex Numbers as Solutions of Quadratic Equations

**Topics**
> The Complex-Number System
> Solutions of Equations

**Key Terms**
Use the vocabulary terms listed below to complete each statement in Exercises 1–3.

|   complex   |   *i*   |   imaginary   |

1. $\sqrt{-1} = $ _____.

2. A(n) _____ number is any number that can be written in the form $a + bi$, where $a$ and $b$ are real numbers and $b \neq 0$.

3. A(n) _____ number is any number that can be written in the form $a + bi$, where $a$ and $b$ are real numbers.

*Express in terms of i.*

4. $\sqrt{-5}$                                    4._____

5. $\sqrt{-49}$                                   5._____

6. $\sqrt{-17}$                                   6._____

7. $\sqrt{-200}$                                  7._____

8. $-\sqrt{-45}$                                  8._____

**9.** $8 - \sqrt{-150}$

**9.**_____

**10.** $\sqrt{-32} - \sqrt{-9}$

**10.**_____

**11.** $2 + \sqrt{-144}$

**11.**_____

**12.** $6 + \sqrt{-81}$

**12.**_____

**13.** $4 - \sqrt{-54}$

**13.**_____

**14.** $7 + \sqrt{-25}$

**14.**_____

**15.** $8 - \sqrt{-18}$

**15.**_____

*Solve*.

**16.** $25x^2 + 4 = 0$

**17.** $16x^2 + 25 = 0$

**18.** $x^2 - x + 5 = 0$

**19.** $(x-6)^2 = -9$

**20.** $(x-5)^2 = -4$

**21.** $x^2 + 40 = 12x$

**22.** $x^2 + 13 = 6x$

**22.**_____

**23.** $x^2 + x + 9 = 0$

**23.**_____

**24.** $2x^2 + 5x + 7 = 0$

**24.**_____

**25.** $2x^2 + 3x + 8 = 0$

**25.**_____

# Chapter 9 QUADRATIC EQUATIONS

### 9.6    Graphs of Quadratic Equations

**Topics**

Graphing Equations of the Form $y = ax^2$

Graphing Equations of the Form $y = ax^2 + bx + c$

**Key Terms**

Use the vocabulary terms listed below to complete each statement in Exercises 1–3

**axis of symmetry**          **parabola**          **vertex**

1.  The graph of a quadratic equation is a(n) _____.

2.  The graph of a quadratic equation is symmetric with respect to its _____.

3.  The maximum or minimum value of a quadratic function occurs at the _____ of its graph.

*Graph each quadratic equation, labeling the vertex and the y-intercept.*

4.   $y = -x^2 + 4$

4.

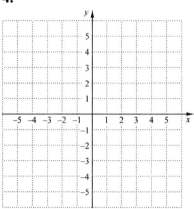

**5.** $y = 2x^2 - 4$

**5.**

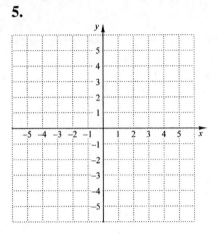

**6.** $y = x^2 + 2x - 2$

**6.**

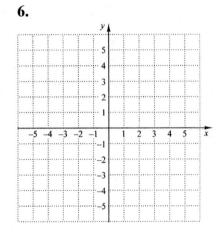

**7.** $y = -x^2 + 4x + 2$

**7.**

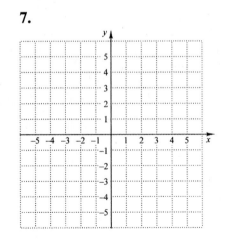

*Graph each quadratic equation, labeling the vertex, the y-intercept, and any x-intercepts. If an x-intercept is irrational, round to three decimal places.*

**8.** $y = 2x^2 - 4x$

**8.**

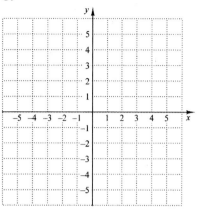

**9.** $y = 2x^2 + 12x + 15$

**9.**

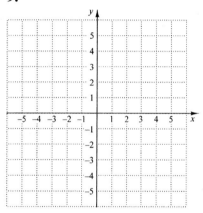

**10.** $y = x^2 - 2x - 5$

**10.**

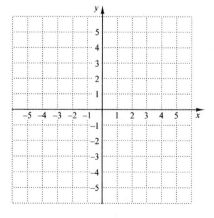

**11.** $y = -2x^2 - 12x - 16$

**11.**

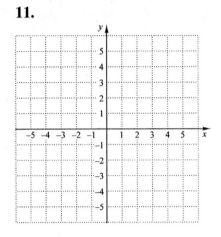

**12.** $y = -x^2 + 6x - 6$

**12.**

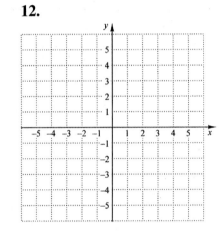

**13.** $y = 2x^2 - 4x + 2$

**13.**

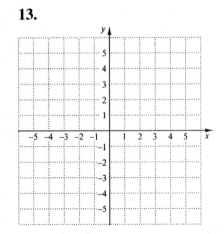

# Chapter 9 QUADRATIC EQUATIONS

## 9.7    Functions

> **Topics**
>> Identifying Functions
>> Functions Written as Formulas
>> Function Notation
>> Graphs of Functions
>> Recognizing Graphs of Functions

**Key Terms**

Use the vocabulary terms listed below to complete each statement in Exercises 1–4.

| **function** | **input** | **independent** | **outputs** |
|---|---|---|---|

1. A(n) _____ is a member of the domain of a function.

2. A _____ is a rule that assigns to each member of some set exactly one member of another set.

3. Function values are also called _____.

4. The variable $x$ in $f(x) = 2x - 7$ is called the _____ variable.

*Determine whether each correspondence is a function.*

5.                                                        5._____

$$
\begin{array}{ll}
-5 & \longrightarrow -4 \\
-3 & \longrightarrow 0 \\
-1 & \longrightarrow 1 \\
2 & \longrightarrow 5 \\
3 &
\end{array}
$$

6.                                                        6._____

$$
\begin{array}{ll}
-5 & \longrightarrow -4 \\
-3 & \longrightarrow 0 \\
-1 & \longrightarrow 3 \\
2 & \longrightarrow 5
\end{array}
$$

*Find the indicated outputs.*

7.  $g(v) = 7v + 1$; find $g(4)$, $g(-2)$, and $g(6.5)$.

7._____

8.  $h(x) = 7x$; find $h(-3)$, $h(6)$, and $h(26)$.

8._____

9.  $g(x) = -2x - 7$; find $g(-2)$.

9._____

10. $f(x) = 3x^2 - 3x$; find $f(0)$, $f(-1)$, and $f(2)$.

10._____

11. $F(x) = |x| - 5$; find $F(15)$, $F(18)$, and $F(-2)$.

11._____

12. $G(x) = z^4 - 1$; find $G(0)$, $G(-1)$, and $G(1)$.

12._____

*Solve.*

13. As the price of a product increases, the consumer's purchase or demand for the product decreases. Suppose that under certain conditions in our economy, the demand for sugar is related to price by the demand function $d(p) = -1.6p + 24.1$, where $p$ is the price of a 5-lb bag of sugar and $d(p)$ is the quantity of 5-lb bags, in millions, purchased at price $p$. Find the quantity purchased when the price is $1 per 5-lb bag.

13._____

*Graph each function.*

**14.**     $f(x) = 3x + 6$

**14.**

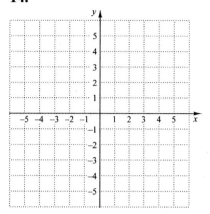

**15.** $g(x) = 4|x|$

**15.**

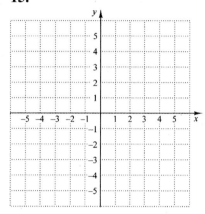

**16.**     $h(x) = 2x^2 - 3$

**16.**

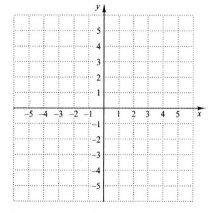

*Determine whether each graph is that of a function.*

**17.**

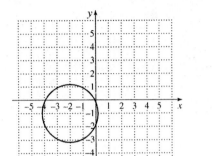

**17.**_____

**18.**

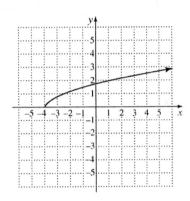

**18.**_____

**19.**

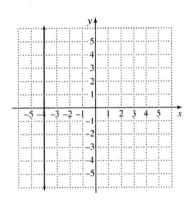

**19.**_____

# Chapter 1 INTRODUCTION TO ALGEBRAIC EXPRESSIONS

## 1.1 Introduction to Algebra

1. constant    3. equation    5. 9    7. 21    9. 372 mi

11. $t + 46$, or $46 + t$    13. $12 + y$    15. $4 \div z$, or $\dfrac{4}{z}$    17. $s + d$, or $d + s$

19. Let $w$ represent "some number"; $\dfrac{1}{2}w$ or $\dfrac{w}{2}$    21. No    23. Yes

25. Let $x$ represent the unknown number; $16x = 128$    27. D    29. A

## 1.2 The Commutative, Associative, and Distributive Laws

1. commutative    3. associative    5. $4x + 9y$    7. $g + uv$

9. $b + (cd + 60)$    11. $(2 \cdot 3)(a + b)$    13. $(65 \cdot s) \cdot r$; $65 \cdot (r \cdot s)$

15. $17y + 102$    17. $28q + 70$    19. $6r + 6n$    21. $g, ghj, 22$

23. $9(x + y)$    25. $6(5r + 7s)$    27. $4, s$    29. $z, 8$

## 1.3 Fraction Notation

1. numerator    3. natural    5. $1 \cdot 175$; $25 \cdot 7$; $35 \cdot 5$; $1, 5, 7, 25, 35, 175$

7. $11 \cdot 17$    9. $\dfrac{5}{6}$    11. $\dfrac{1}{4}$    13. $\dfrac{4}{143}$    15. $\dfrac{7}{27}$    17. $\dfrac{23}{70}$

## 1.4 Positive and Negative Real Numbers

1. irrational    3. inequality    5. $-2100, 4100$    7. $70, -175, 900$

9. 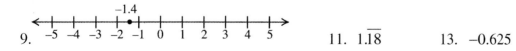    11. $1.\overline{18}$    13. $-0.625$    15. $<$

17. $<$    19. $>$    21. True    23. 26    25. 5.8    27. 0

29. $12, -3$

## 1.5 Addition of Real Numbers

1. D    3. C    5. $-2$    7. 0    9. $-99$    11. 48

283

13. $-110$     15. $-87.5$     17. $-1$     19. $-148$     21. $-26.4$

23. Kyle owes \$407.     25. $9z$     27. $-6b$     29. $19y + 20$

## 1.6 Subtraction of Real Numbers

1. B     3. $c$ minus negative seven     5. $-66$     7. $-12$

9. $-10$     11. $-98$     13. 10     15. $-34$     17. $-12$

19. $-\dfrac{22}{15}$     21. $-\dfrac{13}{6}$     23. $-68-118;\ -186$     25. 155

27. $-4a + 13$     29. $-5a + 6b - 92$

## 1.7 Multiplication and Division of Real Numbers

1. $-2961$     3. 5400     5. 237.8     7. 48.8     9. $-86{,}190$

11. 2     13. Undefined     15. $-\dfrac{10}{7};\dfrac{10}{-7}$     17. $-\dfrac{5}{7}$     19. $\dfrac{1}{1.2}$, or $\dfrac{5}{6}$

21. $\dfrac{10}{9}$     23. $\dfrac{21}{40}$     25. $\dfrac{9}{4}$     27. $-\dfrac{14}{45}$     29. $-2$

## 1.8 Exponential Notation and Order of Operations

1. $t^3$     3. $-125$     5. 12     7. 1     9. 0     11. $-4$

13. $\dfrac{5}{192}$     15. 1     17. 58     19. $-14$     21. $-15$

23. $-29c + 86d - 37$     25. $6z + 29$     27. $-21c + 37d$     29. $n^4 + 2$

# Chapter 2 EQUATIONS, INEQUALITIES, AND PROBLEM SOLVING

## 2.1 Solving Equations

1. 42     3. $-7$     5. 47     7. $\dfrac{41}{24}$     9. $-2.9$     11. 6

13. $-31$     15. $-40$     17. $-\dfrac{5}{8}$     19. $\dfrac{9}{5}$     21. 112.3     23. 62

25. 27     27. 16.9

## 2.2 Using the Principles Together

1. 1     3. −12     5. −7     7. 8     9. 5     11. 5     13. 9

15. 0     17. −22     19. $-\dfrac{9}{5}$     21. $-\dfrac{56}{3}$     23. $\dfrac{9}{7}$     25. $\dfrac{7}{64}$

27. $\dfrac{8}{3}$

## 2.3 Formulas

1. formula     3. 195 mi     5. $240     7. $m = \dfrac{y}{x}$     9. $x = 26 - y$

11. $d^2 = \dfrac{T}{5}$     13. $q = 3A - p - r$     15. $a = Tb$     17. $x = \dfrac{3}{2}(r - 11)$

19. $m = \dfrac{v}{1 + ab}$

## 2.4 Applications with Percent

1. right; write     3. 0.2     5. 0.17     7. 0.78     9. 0.0055

11. 37.2%     13. 91%     15. 62.5%     17. 25%     19. 4     21. 1.3

23. 40%     25. 811     27. 116     29. $45

## 2.5 Problem Solving

1. 21     3. 20 m, 60 m, 120 m     5. 42, 43     7. 34 m, 18 m

9. 215 units     11. 25°

## 2.6 Solving Inequalities

1. a) no     b) yes     c) no     d) no     e) yes

3.     5. $\{x \mid x > -1\}$

7. ![number line with open circle at -3, arrow pointing right]  $-5\ -4\ -3\ -2\ -1\ \ 0\ \ 1\ \ 2\ \ 3\ \ 4\ \ 5$ , $\{x|x>-3\}$

9. $\left\{y\left|y\le\dfrac{3}{14}\right.\right\}$

11. ![number line with open circle at 3, arrow pointing left]  $-5\ -4\ -3\ -2\ -1\ \ 0\ \ 1\ \ 2\ \ 3\ \ 4\ \ 5$ , $\{x|x<3\}$

13. $\left\{x\left|x\ge-\dfrac{8}{9}\right.\right\}$

15. $\{n|n<-3\}$     17. $\{x|x<-5\}$     19. $\{x|x\ge1\}$     21. $\{x|x<6\}$

## 2.7 Solving Applications with Inequalities

1. Let $s$ represent my salary; $s \ge 45{,}000$     3. $80 < s < 120$

5. Let $t$ represent the temperature; $t \le 84$     7. At least 3 hr

9. Scores greater than or equal to 72     11. $2.99

13. A serving of regular cookies contains at least $2\dfrac{2}{3}$ g of fat.

15. At least 4490 minutes     17. At most $4000     19. Widths greater than 33 cm

# Chapter 3 INTRODUCTION TO GRAPHING

## 3.1 Reading Graphs, Plotting Points, and Scaling Graphs

1. $y$-axis     3. origin     5. The executive likely traveled to New York.

7. $1105.28     9. 1425 recordings     11. 2013

13.

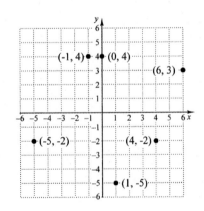

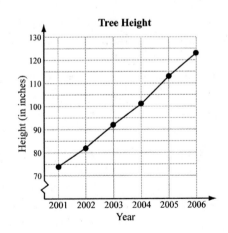

15.

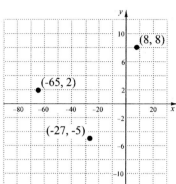

17. II    19. III

## 3.2 Graphing Linear Equations

1.  No    3.    $\dfrac{y=19x-13}{6\mid 19\cdot 1-13}$    $\dfrac{y=19x-13}{25\mid 19\cdot 2-13}$

$6\overset{?}{=}6$  True    $25\overset{?}{=}25$  True

$(-1,-32)$; answers may vary

5.    $\dfrac{6p+5q=10}{6(-5)+5(8)\mid 10}$    $\dfrac{6p+5q=10}{6\cdot 0+5\cdot 2\mid 10}$

$10\overset{?}{=}10$  True    $10\overset{?}{=}10$  True

$(5,-4)$; answers may vary

7.

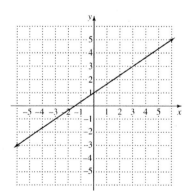

9.

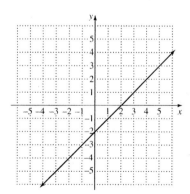

**11.**

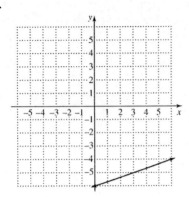

**13.**

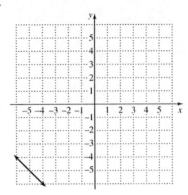

**15.**

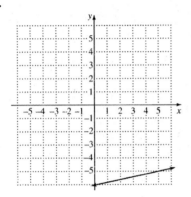

**17.**

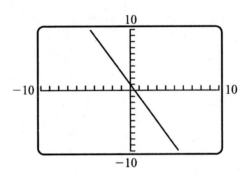

**19.** a

**21.** $400

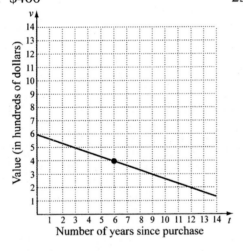

**23.** $800

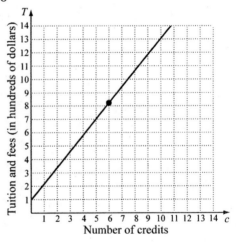

## 3.3 Graphing and Intercepts

1. linear equation     3. vertical line     5. a) $(0, -8)$; b) $(-2, 0), (2, 0)$

7. a) $(0, -3)$; b) $(8, 0)$     9. a) $(0, -8)$; b) none

**11.**

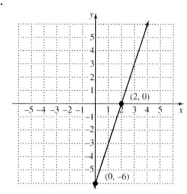

**13.**

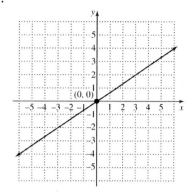

**15.**

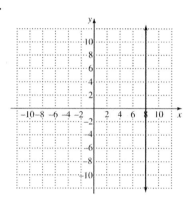

**17.**

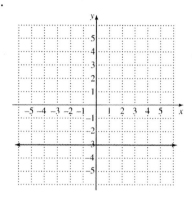

**19.** $y = 1$          **21.** $(-200, 0), (0, 40)$ ; (d)

**23.**

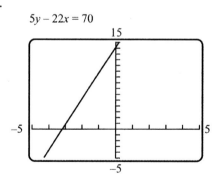

$5y - 22x = 70$

## 3.4 Rates

**1.** a) 18 mi/gal      b) \$40/day   c) 18 mi/day      **3.** a) \$11/hr      b) 6 pages/hr

c) \$1.83/page      **5.** a) 5 streets/min    b) 12 sec/street

7.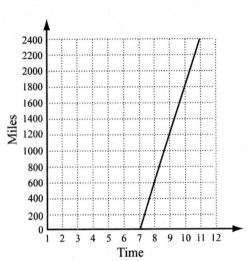

9.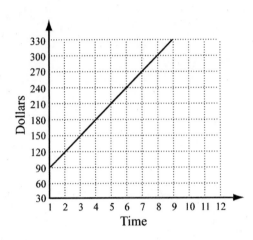

11. $\dfrac{1}{30}$ gal/mi     13. 8¢/min     15. 900 ft/hr     17. d     19. a

## 3.5 Slope

1. 4 million people/yr     3. $2000/yr     5. $\dfrac{4}{3}$     7. 1

9. Undefined     11. $\dfrac{4}{3}$     13. 10     15. $\dfrac{1}{32}$     17. $\dfrac{1}{11}$

19. Undefined     21. 0     23. About 6%     25. About 14%

## 3.6 Slope-Intercept Form

1. slope     3. parallel

5.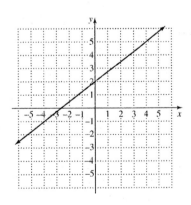

7.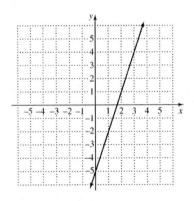

9. $-\frac{3}{5};(0,2)$     11. $\frac{4}{5};(0,-\frac{9}{5})$     13. $y=\frac{2}{9}x+6$     15. $y=-\frac{3}{5}x-\frac{1}{4}$

17. $y = -\frac{1}{15}x + 5$, where $y$ is the distance from home, in miles, and $x$ is the number of minutes spent walking

19.

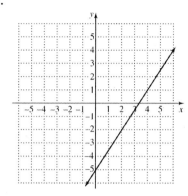

21.

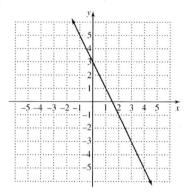

23.

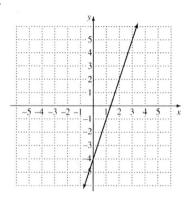

25. No    27. Yes    29. $y = -2x - 3$

31. $y = -\frac{1}{3}x + 2$

## 3.7 Point-Slope Form

1. point-slope form    3. $(x_1, y_1)$    5. extrapolation    7. $y - (-7) = -2(x - 3)$

9. $y - (-8) = \frac{2}{5}(x - (-4))$    11. $-3; (1, -3)$    13. $-\frac{2}{3}; (0, 0)$    15. $y = -\frac{3}{2}x - 6$

17. $y = -\frac{2}{3}x - 8$    19. $y = 10x + 42$

21. 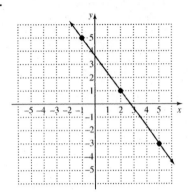      23.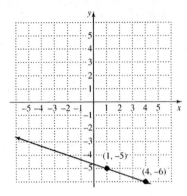

# Chapter 4 POLYNOMIALS

## 4.1 Exponents and Their Properties

1. $u^{13}$    3. $(5s)^{16}$    5. $(x+3)^7$    7. $u^4$    9. $27s^2$    11. $x^6 y^3$

13. 1    15. 11    17. $8c^3$    19. $625q^{36}$    21. $81x^{60}y^{80}$    23. $\dfrac{27}{343m^3}$

25. $\dfrac{n^4}{64}$    27. $\dfrac{a^{16}}{81b^{12}}$

## 4.2 Polynomials

1. polynomial    3. degree    5. No    7. Coefficients: 5, 6, 3;

degrees: 2, 1, 0    9. Trinomial    11. $x^6 + x^5$    13. $-9x^5 + 8x$

15. 10; 26    17. 15; 27    19. 16 ft    21. About 200 mg

## 4.3 Addition and Subtraction of Polynomials

1. $-x+7$    3. $-1.9x^3 - 0.3x^2 - 4.5x + 62$    5. $-\dfrac{1}{2}x^4 + \dfrac{1}{4}x^3 + x^2$

7. $-6z^8 + 4z^7 + 11z$    9. $-4x^2 + 4x - 7$    11. $-x^2 - 7x + 17$

13. $5.8x^3 + 8.4x^2 - 3.3x - 92$    15. $\dfrac{1}{3}x^3 - \dfrac{3}{4}x$    17. $4x+5$

19. $(w+12)(w+9)$; $w^2 + 21w + 108$    21. $(w^2 - 9\pi)$ ft$^2$

## 4.4 Multiplication of Polynomials

1. $12x^9$     3. $0.16x^{10}$     5. $0$     7. $-2x^2 + 14x$     9. $x^8 + x^3$

11. $-4x^{10} - 4x^6$     13. $x^2 + 6x + 8$     15. $x^2 - 9x + 8$     17. $\dfrac{12}{49}a^2 + a - 12$

19.

$x + 3 \begin{cases} \phantom{x} \\ \phantom{x} \end{cases}$

| | $3$ | $3x$ | $18$ |
| $x$ | $x^2$ | $6x$ |

$\underbrace{\phantom{xxxxxxxx}}_{x + 6}$  $x$   $6$

21. $8x^3 + 16x^2 + 10x + 2$

23. $-40x^5 - 17x^4 + 5x^3 + 10x^2 - 2x$     25. $54t^4 + 27t^3 - 66t^2 - 35t + 6$

## 4.5 Special Products

1. $x^9 + 9x^8 + 4x + 36$     3. $36x^2 + 48x + 16$     5. $4x^2 - 42x + 54$

7. $-8x^2 - 49x + 49$     9. $45x^8 + 25x^6 + 63x^2 + 35$     11. $36x^{18} - 9$

13. $x^2 + 16x + 64$     15. $a^2 - 9a + \dfrac{81}{4}$     17. $x^4 + x^3 - 5x^2 - 4x + 4$

19. $9x^3 + 63x^2 - 45x$     21. $25x^6 - 90x^3 + 81$     23. $49x^{14} + 84x^7 + 36$

25. $x^2 + 8x + 16$     27. $9x^2 + 21x + 10$     29.

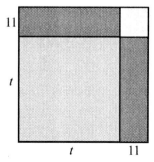

Answers to Worksheets for Classroom or Lab Practice

## 4.6 Polynomials in Several Variables

1. 7     3. 3.38 L     5. 170.4 m     7. Coefficients: $22, -6, -9$;

degrees: 12, 6, 0; 12     9. $3x^7 y - 2xy^7 + x^3$     11. $7x^2 - 7xy + 3y^2$

13. $-2x^2 - 6xy - 9y^2$     15. $a^{10}b^2 - 13a^5 b + 42$     17. $25x^2 + 60xh + 36h^2$

19. $v^4 - t^2 j^2$     21. $x^2 - y^2 - 2y - 1$     23. $s^2 + 2st + t^2$

25. $r^2 + s^2 + t^2 + 2rs + 2rt + 2st$     27.

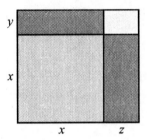

## 4.7 Division of Polynomials

1. quotient     3. dividend     5. $3a^8 - 8a^3$     7. $2x^4 - 10x^3 + 4$

9. $6x^4 - 10x^2 - \dfrac{9}{2}$     11. $x + 8$     13. $a^2 - a - 1 + \dfrac{4}{a+1}$     15. $3x - 2$

17. $2x^2 + 3x + 1 + \dfrac{8x-3}{x^2-3}$     19. $x^2 - x - 1 + \dfrac{-4}{x-2}$     21. $a^2 - 8a + 28 + \dfrac{-87}{a+3}$

23. $t^3 + 2t^2 + 4t + 8$

## 4.8 Negative Exponents and Scientific Notation

1. $\dfrac{256}{81}$     3. $r^t$     5. $y^{-2}$     7. $5^{-1}$     9. $s^{-13} = \dfrac{1}{s^{13}}$     11. $x^{13}$

13. $\dfrac{t^6 w^{30}}{s^{18} p^6}$     15. 1     17. $4.8 \times 10^{-2}$     19. $1.31 \times 10^{-1}$

# Chapter 5 POLYNOMIALS AND FACTORING

## 5.1 Introduction to Factoring

1. factor      3. largest common factor      5. Answers may vary. $-7a^9 \cdot 3a$,

$21a^5 \cdot -a^5$, $-3a^2 \cdot 7a^8$      7. $4t^2(t-5)$      9. $7(a^2-2a+7)$

11. $-10(x-4)$      13. $3(a-b)$      15. $4m^2(5m^2-6m-7)$      17. $7x^5\left(x^2-7\right)$

19. $(y+2)(x+z)$      21. $(t-1)(6t^3+5)$      23. $(x-5)(x^2-2)$      25. $(5-p)(pq-4)$

## 5.2 Factoring Trinomials of the Type $x^2 + bx + c$

1. trinomial      3. prime      5. constant term      7. $(x+2)(x+9)$

9. $x(x+2)(x-6)$      11. $(a-4)(a-11)$      13. $(x+3)(x+12)$

15. $(x+6)(x-3)$      17. $(a+2b)(a-11b)$      19. $(w+5)(w+3)$

21. $(s-8)(s-4)$      23. $3(s-3)(s-7)$      25. $-b(b+8)(b-9)$

## 5.3 Factoring Trinomials of the Type $ax^2 + bx + c$

1. $(5x+3)(2x-1)$      3. $(10t+3)(4t-5)$      5. Prime

7. $-2(3z+1)(3z-5)$      9. $ab(4a+5)(2a-3)$      11. $3t^2(5t+2)(2t-5)$

13. $(5c-2d)(5c-2d)$, or $(5c-2d)^2$      15. $(v+5)(v+6)$      17. $(2s+3)(3s+2)$

19. $(5b+3)(6b-7)$      21. $3(3m-4)(m+2)$      23. $(28b-1)(b+4)$

25. $(7c-9v)(8c+7v)$

## 5.4 Factoring Perfect-Square Trinomials and Differences of Squares

1. factored completely      3. perfect-square trinomial      5. No      7. No

9. $(y+4)^2$      11. $(5-n)^2$, or $(n-5)^2$      13. $x(x+10)^2$      15. Yes

17. No      19. $(3+mn)(3-mn)$      21. $15(x^2+y^2)(x+y)(x-y)$

23. $\left(\frac{1}{2}+y\right)\left(\frac{1}{2}-y\right)$    25. $(x+3)(x-3)(x+1)$    27. $2(2s+7)(2s-7)$

29. $(3s+4f)(3s-4f)$

## 5.5 Factoring: A General Strategy

1. common factor    3. grouping    5. $(2x+3)(x-2)$    7. $a(a-6)^2$

9. $(x-2)(x+3)(x-3)$    11. Prime    13. $4(x-7)(3x+1)$

15. $(c^2+5)(c^2-5)$    17. $c^2(x+y)$    19. $(5x-y)(2x-1)$

21. $(t-2+10v)(t-2-10v)$    23. $(3p-2q)^2$    25. Prime

27. $(4m-3n)(5m+2n)$    29. $(a^2+b^3+5)(a^2-b^3-5)$    31. $\left(\frac{1}{3}a+\frac{2}{5}\right)^2$

## 5.6 Solving Quadratic Equations by Factoring

1. $1, 9$    3. $0, 9$    5. $-3, 0, 2$    7. $-25, 2$    9. $-2, 0$    11. $-\dfrac{5}{3}, 1$

13. $-\dfrac{5}{2}, \dfrac{3}{8}$    15. $6$    17. $-8, 8$    19. $-3, 1$

## 5.7 Solving Applications

1. right triangle    3. leg    5. $-16, 15$    7. 2 cm    9. $-11, -9, -7$ or

7,9,11    11. Height: 8 ft; base: 6 ft    13. 20 ft    15. 4 lamps    17. 5 yr

# Chapter 6 RATIONAL EXPRESSIONS AND EQUATIONS

## 6.1 Rational Expressions

1. simplify    3. $-\dfrac{5}{9}$    5. $4, -4$    7. $\dfrac{3}{4r^2}$    9. $a-2$

11. $\dfrac{q+1}{7q+5}$    13. $\dfrac{4}{t+3}$    15. $\dfrac{n+1}{7n+6}$    17. $-1$    19. $x+5$    21. $\dfrac{5}{6}$

23. $\dfrac{x-6}{x+2}$

## 6.2 Multiplication and Division

1. reciprocal

3. $\dfrac{2a(a+1)}{7(5a-3)}$

5. $\dfrac{6}{x^5}$

7. $\dfrac{8}{a-b}$

9. $4t(t+1)$

11. $\dfrac{1+3y}{2(y-1)}$

13. $\dfrac{11x}{10}$

15. $\dfrac{24}{x^2}$

17. $\dfrac{3(x+1)(x-1)}{x-7}$

19. $\dfrac{(x+2)(x+3)}{3(x-5)}$

21. $\dfrac{x+2}{x^5}$

## 6.3 Addition, Subtraction, and Least Common Denominators

1. $\dfrac{10}{z}$

3. $\dfrac{d+11}{d}$

5. $y+11$

7. $t+7$

9. $\dfrac{a+1}{a+4}$

11. $\dfrac{x+18}{x-9}$

13. $\dfrac{3z+1}{z^2-5z-36}$

15. $72$

17. $50x^8$

19. $55(z-11)$

21. $x^7(x-2)^2(x^2+4)$

23. $(x-3)(x+7)^2$

25. $\dfrac{44}{20x^6},\ \dfrac{x^2y}{20x^6}$

27. $\dfrac{(t+3)(t+5)}{(t+4)(t-4)(t+5)},\ \dfrac{(t-4)^2}{(t+4)(t-4)(t+5)}$

## 6.4 Addition and Subtraction with Unlike Denominators

1. $\dfrac{9v+4}{v^2}$

3. $\dfrac{8y+5t}{y^2t^2}$

5. $\dfrac{9t+47}{56}$

7. $\dfrac{3w-32}{11w}$

9. $\dfrac{2v-40}{(v+4)(v-4)}$

11. $\dfrac{-6x+25}{x^2-9}$

13. $\dfrac{3v-17}{(v-8)^2}$

15. $\dfrac{w-6}{(w+4)(w+6)}$

17. $\dfrac{7x-6}{(x-6)(x+6)^2}$

19. $\dfrac{p^2+8p+8}{(p-3)(p+3)}$

21. $\dfrac{2a-4}{(a-4)(a+4)}$

23. $\dfrac{4z-1}{z-5}$

25. $\dfrac{-2w^2-2w-7}{(w+1)(w-1)}$

## 6.5 Complex Rational Expressions

1. $\dfrac{(x-2)(x-5)}{(x+3)(x+1)}$

3. $z(y-z)$

5. $-x-y$

7. $\dfrac{(x+1)(x+3)}{(x+2)(x+5)}$

9. $\dfrac{1-6x}{11x+1}$  11. $\dfrac{x^2-2x+6}{x^2+2}$  13. $\dfrac{x+5}{x}$  15. $\dfrac{-x-4}{x-2}$

### 6.6 Solving Rational Equations

1. rational equation   3. LCD   5. 20   7. −6   9. 2, 5

11. 3   13. −2, 7   15. 4   17. 7   19. No solution

### 6.7 Applications Using Rational Equations and Proportions

1. $\dfrac{t}{a}+\dfrac{t}{b}=1$   3. $\dfrac{10}{7}$   5. −11, −10 or 10, 11   7. $9\dfrac{3}{13}$ hr

9. Alex: 15 min; Ryan: 35 min   11. 8 hr   13. Scooter: 42 mph;

motorcycle: 58 mph   15. 4 km/h   17. 2.4   19. 2.4 in.

21. 96 pages   23. 45 trout

# Chapter 7 SYSTEMS AND MORE GRAPHING

### 7.1 Systems of Equations and Graphing

1. system   3. consistent   5. dependent   7. No   9. $(4,3)$

11. No solution   13. $\{(x,y)\,|\,y=2x+1\}$   15. Exercises 9, 10, 12, 13, and 14

### 7.2 Systems of Equations and Substitution

1. $(2,3)$   3. $(1,1)$   5. Infinite number of solutions   7. $(\frac{3}{4},1)$

9. $(-3,-12)$   11. No solution   13. $(-8,-5)$   15. 35, 42

17. 36, 20   19. $43°, 47°$   21. Length: 144 m; width: 70 m

### 7.3 Systems of Equations and Elimination

1. $(7,4)$   3. $(-3,-12)$   5. $(\frac{2}{5},-\frac{16}{5})$   7. No solution

9. Infinite number of solutions   11. $(40,-120)$   13. 49.8 mi

15. $17\frac{21}{22}$ min   17. Brittany: 520 awards; Jon: 320 awards

## 7.4 More Applications Using Systems

1. distance, rate, time    3. Sara: 53 mi; Jean: 43 mi

5. Children's plates: 185; adults' plates: 140    7. 16-oz candles: 34;

24-oz candles: 48    9. Alpine: $4\frac{2}{3}$ lb; Meadows: $5\frac{1}{3}$ lb    11. Tropical Punch:

12 L; Caribbean Spring: 6 L    13. Graphics: 146; ringtones: 80

## 7.5 Linear Inequalities in Two Variables

1. linear inequality    3. Yes

5. 

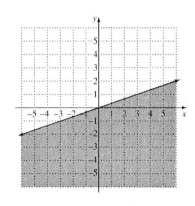

7.

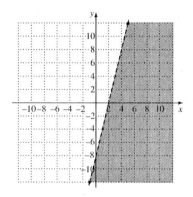

9.

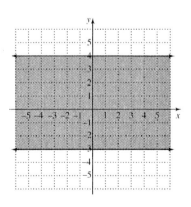

11.

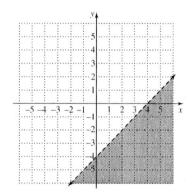

13.

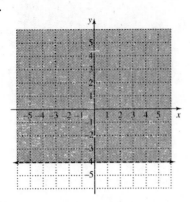

## 7.6 Systems of Linear Inequalities

1.

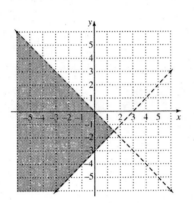

3.

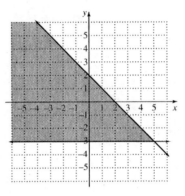

5.

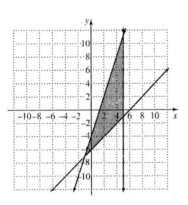

7.

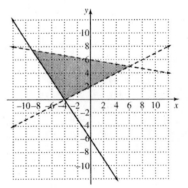

9.

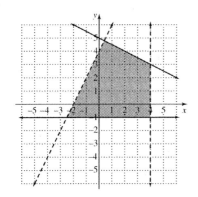

## 7.7 Direct and Inverse Variation

1. direct 3. $y = 11x$ 5. $y = \dfrac{63}{x}$ 7. $16\frac{2}{3}$ cm 9. 3.6 hr

11. 120 m 13. $y = \frac{5}{2}x^2$ 15. 3.125 W/m²

# Chapter 8 RADICAL EXPRESSIONS AND EQUATIONS

## 8.1 Introduction to Square Roots and Radical Expressions

1. principal 3. radicand 5. False 7. $8, -8$ 9. $-8$

11. 0.3 13. 20 15. $x + 1$ 17. Rational 19. 1.414 21. $2|x|$

23. $|x + 7|$ 25. $31d$ 27a. 10 b. 14

## 8.2 Multiplying and Simplifying Radical Expressions

1. $\sqrt{110}$ 3. $\sqrt{\dfrac{3}{10}}$ 5. $7\sqrt{5}$ 7. $5\sqrt{2pq}$ 9. $3\sqrt{10}$

11. $6\sqrt{x}$ 13. $5x$ 15. $2\sqrt{46}$ 17. $x^2$ 19. $t^4\sqrt{t}$ 21. $7\sqrt{6}$

23. $15\sqrt{xy}$ 25. $a\sqrt{bc}$ 27. $28x^5$ 29. 63.2 mph; 24.5 mph

## 8.3 Quotients Involving Square Roots

1. rationalizing 3. $\dfrac{9}{7}$ 5. $\sqrt{2}$ 7. $5y$ 9. $3d^6$ 11. $\dfrac{5\sqrt{3}}{3}$

13. $\dfrac{2\sqrt{5}}{25}$  15. $\dfrac{3\sqrt{14}}{10}$  17. $\dfrac{\sqrt{3y}}{6}$  19. $a^2\sqrt{3b}$

21. 7.85 sec; 3.14 sec

## 8.4 Radical Expressions with Several Terms

1. like radicals  3. $-3\sqrt{11}$  5. $2\sqrt{7}$  7. $8\sqrt{2}$  9. $\sqrt{7}$

11. $11\sqrt{2}$  13. $6-2\sqrt{30}$  15. $16+4\sqrt{5}$  17. $2x+7-2\sqrt{14x}$

19. $\dfrac{5-\sqrt{3}}{11}$  21. $\dfrac{p+\sqrt{pq}}{p-q}$  23. $\dfrac{6\sqrt{14}-6}{13}$  25. $\dfrac{7+\sqrt{14}}{5}$

27. $\dfrac{3\sqrt{19}+3\sqrt{2}}{17}$  29. $-12-6\sqrt{5}-2\sqrt{7}-\sqrt{35}$

## 8.5 Radical Equations

1. radical equation  3. 64  5. $\dfrac{50}{3}$  7. 14  9. 0, 49  11. 5

13. No solution  15. 2, 6  17. 2  19. 15  21. $-15$  23. 1

25. 16 m

## 8.6 Applications Using Right Triangles

1. $3\sqrt{2}$; 4.243  3. 4  5. $\sqrt{65}$; 8.062  7. 15  9. 4 m

11. 5 cm  13. $5\sqrt{29}$ ft; 26.926 ft  15. $2\sqrt{22}$ ft; 9.381 ft

17. $\sqrt{221}$ m; 14.866 m  19. $\sqrt{85}\approx 9.220$  21. $3\sqrt{2}\approx 4.243$

## 8.7 Higher Roots and Rational Exponents

1. b  3. c  5. cube  7. $-\dfrac{2}{3}$  9. $-5$  11. $-2t$

13. $\dfrac{10x^2\sqrt[3]{x}}{3y}$  15. $\dfrac{mn}{p^3}\sqrt[6]{\dfrac{m}{p^2}}$  17. 32  19. 7  21. 27

23. 243    25. $\dfrac{1}{9}$    27. $\dfrac{1}{8}$    29. $\dfrac{1}{16}$

# Chapter 9 QUADRATIC EQUATIONS

## 9.1 Solving Quadratic Equations: The Principle of Square Roots

1. $-14, 14$    3. $-\sqrt{2}, \sqrt{2}$    5. $-\sqrt{3}, \sqrt{3}$    7. $-\dfrac{7}{9}, \dfrac{7}{9}$

9. $-\dfrac{\sqrt{34}}{8}, \dfrac{\sqrt{34}}{8}$    11. $-\dfrac{\sqrt{21}}{3}, \dfrac{\sqrt{21}}{3}$    13. $7, 3$

15. $-5+\sqrt{6}, -5-\sqrt{6}$    17. $2, 16$    19. $-7, 3$

21. $\dfrac{2}{7} \pm \dfrac{\sqrt{35}}{7}$, or $\dfrac{2 \pm \sqrt{35}}{7}$    23. $-10, -4$    25. $-6, 2$

## 9.2 Solving Quadratic Equations: Completing the Square

1. $x^2 + 20x + 100 = (x+10)^2$    3. $t^2 - 5t + \frac{25}{4} = (t - \frac{5}{2})^2$

5. $81; (t-9)^2 = t^2 - 18t + 81$    7. $\dfrac{49}{4}; \left(t - \dfrac{7}{2}\right)^2 = t^2 - 7t + \dfrac{49}{4}$

9. $\dfrac{25}{144}; \left(n - \dfrac{5}{12}\right)^2 = n^2 - \dfrac{5}{6}n + \dfrac{25}{144}$    11. $2 \pm \sqrt{2}$    13. $-6, 1$

15. $9 \pm \sqrt{85}$    17. $-1 \pm \sqrt{5}$    19. $-5, 4$    21. $\dfrac{-13 \pm \sqrt{377}}{4}$, or

$-\dfrac{13}{4} \pm \dfrac{\sqrt{377}}{4}$    23. $\dfrac{-5 \pm \sqrt{33}}{4}$, or $-\dfrac{5}{4} \pm \dfrac{\sqrt{33}}{4}$    25. $\dfrac{1}{2}, \dfrac{3}{4}$

## 9.3 The Quadratic Formula and Applications

1. $\pm\sqrt{k}$    3. $-\dfrac{3}{2} \pm \dfrac{\sqrt{17}}{2}$    5. No real-number solution

7. $-2 \pm \dfrac{3\sqrt{2}}{2}$    9. $-\dfrac{1}{4} \pm \dfrac{\sqrt{105}}{20}$    11. $1, 3$    13. No real-number solution

15. $\dfrac{9\pm\sqrt{421}}{10}$    17. $-3.449, 1.449$    19. $-3.121, 1.121$

21. 8 sides    23. 15 ft, 36 ft    25. Length: 12 ft; width: 5 ft    27. 18.8%

## 9.4  Formulas

1. $v=\dfrac{9A}{t}$    3. $C=\dfrac{P^2}{196}$    5. $A=\dfrac{-u+\sqrt{u^2+4vS}}{2v}$

7. $v=\sqrt{\dfrac{T}{8}}$, or $v=\dfrac{\sqrt{2T}}{4}$    9. $w=\sqrt{\dfrac{kxz}{y}}$, or $w=\dfrac{\sqrt{kxzy}}{y}$

11. $c=\sqrt{a^2-b^2+2abX}$    13. $x=1+\sqrt{1+3B}$    15. $f=\dfrac{e}{M-C}$    17. $\pm\sqrt{wr}$

19. $K=\dfrac{D}{N}$    21. $a=\sqrt{d^2+f^2}$

## 9.5  Complex Numbers as Solutions of Quadratic Equations

1. $i$    3. complex    5. $7i$    7. $10i\sqrt{2}$, or $10\sqrt{2}i$

9. $8-5i\sqrt{6}$, or $8-5\sqrt{6}i$    11. $2+12i$    13. $4-3\sqrt{6}i$    15. $8-3\sqrt{2}i$

17. $\pm\dfrac{5i}{4}$    19. $6\pm 3i$    21. $6\pm 2i$    23. $-\dfrac{1}{2}\pm\dfrac{\sqrt{35}}{2}i$    25. $-\dfrac{3}{4}\pm\dfrac{\sqrt{55}}{4}i$

## 9.6  Graphs of Quadratic Equations

1. parabola    3. vertex

5.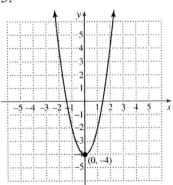

7.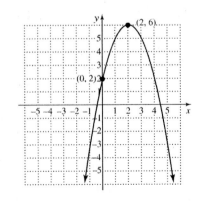

9.

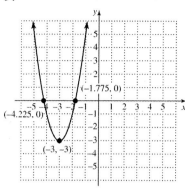

11.

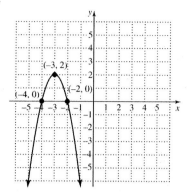

13.

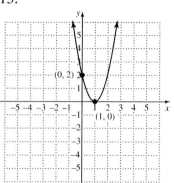

## 9.7 Functions

1. input     3. outputs     5. Yes     7. 29; −13; 46.5     9. −3

11. 10; 13; −3     13. 22.5 million bags

15.

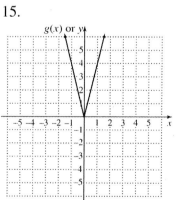

17. No     19. No